BINOCULAR VISION

BINOCULAR VISION

The Politics of Representation in Birdwatching Field Guides

SPENCER SCHAFFNER

UNIVERSITY OF MASSACHUSETTS PRESS
Amherst and Boston

Printed in the United States of America

LC 2011015729
ISBN 978-1-55849-886-0 (paper); 885-3 (library cloth)

Designed by Sally Nichols
Set in Monotype Bell and Priori Sans
Printed and bound by Thomson-Shore, Inc.

Library of Congress Cataloging-in-Publication Data

Schaffner, Spencer, 1970-
Binocular vision : the politics of representation in birdwatching field guides /
Spencer Schaffner.
p. cm.
Includes bibliographical references and index.
ISBN 978-1-55849-886-0 (pbk. : alk. paper) —
ISBN 978-1-55849-885-3 (library cloth : alk. paper)
1. Bird watching—Political aspects. I. Title.
QL677.5.S33 2011
598.072'34—dc23

2011015729

British Library Cataloguing in Publication data are available.

Publication of illustrations has been made possible by the
generous support of the Campus Research Board at the University of Illinois,
Urbana-Champaign

In loving memory of my father, who taught me how to watch birds

CONTENTS

ILLUSTRATIONS

ACKNOWLEDGMENTS

This book could not have been written without the support of family, friends, colleagues, and institutions. I am particularly thankful for the support from my friends and colleagues in the English department and Center for Writing Studies at the University of Illinois, Urbana-Champaign.

When new field guides are published, old ones can seem obsolete, so I am indebted to the many libraries and librarians who have maintained diverse, regional collections of field guides. I had the privilege of examining well-maintained print field guides at the University of Washington Libraries, Stratton Library at Sheldon Jackson College, the University of Utah J. Willard Marriot Library, the Penn State University Libraries, and the University Library at the University of Illinois, Urbana-Champaign. In particular, I would like to thank librarian Sandra Kruppa, in Special Collections at the University of Washington, for taking me under her wing and teaching me how to read with a loupe.

Although their identities must remain anonymous, I am equally indebted to the 237 Pacific Northwest birdwatchers who volunteered to participate in my qualitative field research. These interactions with birdwatchers challenged many of my preconceived notions. Members of TWEETERS and OBOL, I thank you.

I am also indebted to a number of editors who have encouraged my work on this topic: Monica Casper, C. L. Cole, Richard Crombet-Beolens, Sid Dobrin, Gordon Hutner, Diana Mincyte, Sean Morey, and Rick Wright. At the University of Massachusetts Press, Paul Wright expressed an early and encouraging interest in this book, and Brian Halley and Bruce Wilcox were extremely helpful in moving the project to completion. At the University of Illinois, I am indebted to colleagues John Marsh, Justine Murrison, Debra Hawhee, and Catherine Prendergast. Melissa Littlefield read and reread several drafts of the book, offering invaluable suggestions. Phineas Reichert kindly joked with me through many stages of the project. I also benefited from the thorough and candid suggestions of multiple anonymous reviewers.

Completion of this project was made possible by generous funding from the Illinois Campus Research Board and the English Department at the University of Illinois, Urbana-Champaign. A small portion of chapter 1 previously appeared in *American Literary History* and is reprinted here with permission from Oxford University Press. An earlier version of chapter 5 was published in the *Journal of Sport and Social Issues* and is reprinted here with the permission of Sage. No one who helped me complete this book necessarily shares the views expressed in these pages, and all shortcomings are my own.

BINOCULAR VISION

INTRODUCTION

Field guides are enchanting, visual texts. With brilliant illustrations, careful design, and terse scientific descriptions, field guides help answer a fundamental question: *What am I looking at?* With the help of a field guide, an unknown tree can become a Sugar Maple (*Acer saccharum*) or a hard-to-identify raptor a Ferruginous Hawk (*Buteo regalis*). Field guides help render the things around us recognizable, classifiable, and predictable.

With their attention to taxonomy, field guides typically ignore connections between categories—links between trees and birds, for instance—as well as human alterations to the environment at large. In the case of birdwatching field guides, the books feature easily identifiable, healthy-looking birds in unadulterated surroundings. In recent years, though, several unusual field guides have challenged some of the basic assumptions and political implications of more mainstream guides. *A Field Guide to Common Birds of Toronto,* for example, features species that are frequently killed from flying into Toronto-area windows (MacKay, Riss, and Antonello 2009). The illustrations in this guide are all of dead birds, and the species descriptions include phrases such as "usually dies during the night due to window collisions" or "found dead or severely injured at the base of office buildings."

The wildlife photographer Chris Jordan has created another haunting set of images that make up a sort of field guide, this one of photographs of albatross chicks that died on Midway Atoll after ingesting large quantities of plastics. The goal of a typical field guide is to render each bird distinguishable as a member of a particular species, but the birds in Jordan's "Midway: Message from Gyre" (2009) photo essay are hard to tell apart given their advanced states of decomposition. In another unusual field guide, William Tice's *A Birder's Guide to the Sewage Ponds of Oregon: Creatures from the Brown Lagoons* (1999), readers are instructed on how to find birds at sewage treatment facilities. Whereas field guides typically represent birds in bucolic, even hygienic natural settings, Tice presents the sewage pond as an overlooked birdwatching treasure.

These unusual guides suggest that how field guides answer the question *What am I looking at?* has political and environmental consequences. Throughout *Binocular Vision: The Politics of Representation in Birdwatching Field Guides*, I pay attention to the basic assumptions and implications of more mainstream guides. Field guides not only represent birds but put forth an entire worldview about the environments they are part of. With an estimated 40 million to 70 million birdwatchers in North America,[1] field guides are not only profitable but also political: several mainstream environmental groups endorse and publish their own guides. The National Audubon Society, for instance, has several guides, including David Allen Sibley's best-selling *The Sibley Guide to Birds* (2000), which sold 500,000 copies in its first two years of circulation (Cordell and Herbert 2002, 54).[2] Another popular guide, the National Geographic Society's *Field Guide to the Birds of North America* (Allen and Hottenstein 1983, 1987, 1999, 2002), has sold more than 1.5 million copies in four editions, raising funds for that organization.[3] The sale of these texts generates revenue for the groups that endorse and publish them, but that is not all that results from this transaction: the marketing of field guides in this way assumes and establishes connections between field guides, birdwatching, and environmental conservation.

These connections between guides, practices of birdwatching, and larger environmental ideologies are not necessarily stable, however. Take Robert Elman's *The Hunter's Field Guide to the Game Birds and Animals of North America* (1974), for example. Field guides for hunters are meant to help users identify birds, and hunting groups have long been integral to the conservation movement (Philippon 2004, 33–71), but hunting field guides do not produce the same kinds of associations between hunting and mainstream environmental conservation that birdwatching field guides do. Bird guides, as opposed to hunting guides, appear "green."

Despite the linkages between the field guides birdwatchers use, birdwatching, and conservation which have been established in part by the nonprofit groups that sponsor birdwatching field guides, I insist throughout this book that the forms of understanding that field guides promote are not necessarily conservationist. Instead, field-guide authors represent birds and the environments they live in as strangely detached from and unaffected by a wide array of currently pressing environmental challenges.[4] In addition, as widely circulating interpretive tools used for understanding the natural world, field guides promote an equally problematic approach to examining, interpreting, and understanding "nature."

In *Binocular Vision* I explore the ideological workings of field guides, suggesting that, above all, field guides are produced by and productive of singular, narrow, distinguishing, and decontextualizing ways of seeing, approaching, and thinking about the natural world. Instead of representing birds as enmeshed in multidimensional networks including environmental issues and problems, mainstream field guides sanitize representations of birds and the worlds they live in, representing birds as living apart from one another and from the consequences of human intervention. In this way, field guides are strong promoters of what I call *binocular vision*, a taxonomic, focused way of seeing and thinking about individual parts of "nature" as disconnected from one another and from humans. Field guides, as I describe them, are political texts that are involved in much more than simple acts of identification. As Jeffrey Karnicky has argued, birdwatching is not merely an outdoor hobby but a complex "human visual apparatus" (2004, 254), and one of the elements in that apparatus is the field guide.

By questioning key methods of representation in field guides, this book challenges the common belief that knowing more about such things as insects, plants, and animals is a vital first step toward becoming a conservationist. This belief helps sustain a sense that the connections between field guides and environmental conservation are beyond question. Roger Tory Peterson, the most well known and prolific field-guide author of the twentieth century, strongly endorsed this view, saying, "Make a birder, and you make a conservationist" (Rosen 2000, 3; Clement 2003, 915). In large part, Peterson's guides were intended to foster an increasingly aware populace of conservationists.

Although there is no doubt that birdwatchers have contributed actively to environmental conservation on many levels (Barrow 2002) and that vital conservationist legislation such as the Migratory Bird Treaty Act of 1918 and the Endangered Species Act of 1973 have been supported by many birdwatchers, field guides are less than ideal do-it-yourself, conservation-start-up kits. Indeed, they are fraught with limitations, and to the extent that texts are instrumental in the formation of social consciousness, field guides ignore and efface critical aspects of a robust conservationism. Most birdwatchers use field guides and many birdwatchers are conservationists, but clear causal relationships between field guides, birdwatching, and conservationism are based on shaky suppositions.

To get at these sometimes problematic underlying assumptions behind field guides, I discuss a wide range of guides, comparing them to one another historically while bringing in similarly technical representations of birds in other domains such as art, advertising, and environmental law.[5] In showing how the texts present *perspectives* on birds and the environment that are anything but objective or transparent, I discuss three main assumptions. One, which is most apparent in early field guides, is the sense that there are good and bad, or exotic and undesirable, birds. Judgments of this kind have changed over time, but early field guides from the end of the nineteenth century make it appear as if such distinctions can be unambiguously determined. As my discussions of species such as the Bald Eagle and Mute Swan show (these are species whose reputations as "good" and "bad" birds have changed dramatically), values associated with different birds are tied up in cultural debates about such things as environmental nationalism and biodiversity. Overt judgments about birds largely drop out of the more technical twentieth-century guides, although sentimental hierarchies of preference continue to inform debates over bird bounties, species conservation, and the ongoing lethal management of nuisance birds.

Representations of birds as unambiguously good or bad, exotic or common, is connected to the second, related assumption I examine, which is that birds can be conclusively studied visually via a system of field marks. As I explain, this visual approach to bird study developed in the ways that it did in field-guide literature in part because of the affordances and limitations of the printed book: the printed page has proven highly amenable to representing how birds look and less amenable to representing how they sound. Authors have struggled with an array of methods for representing bird sounds with fidelity, each of them having numerous problems, while multicolor offset printing has allowed for the production of increasingly vivid and colorful field-guide images. In recent years, this long-standing assumption that birds in particular and the environment more broadly are best studied visually has been highlighted by a new array of electronic field guides that, although still quite visual, feature a new capacity to store playable audio files. These new guides encourage a renewed interest in bird sounds and a realization that bird*watching* is but one way to approach nature study. The visual disposition that field guides promote is important in that it proves inherently limited in the examination of largely invisible environmental factors such as toxic pollution. Binocular vision, in this sense, is myopic.

The third assumption I examine is one I alluded to already: the sense that the "nature" that birds are part of is entirely disconnected from human influences. Even though human technologies and myriad environmental modifications play an increasing role in the lives of birds, few field guides recognize the extent to which the habitats birds live in are altered and that those alterations greatly affect bird life. As I describe, this lacuna is in part because field-guide illustration borrows heavily from the representational traditions of zoological illustration. These forces are not insurmountable, however. One radical field-guide author, Jack Griggs, counters these visual traditions by featuring images of birds living in highly modified landscapes.

With so many powerful assumptions built into their pages—about good and bad birds, a visual approach to studying nature, and the separation of birds and humans—field-guide literature represents birds as unencumbered by a range of pressing environmental problems. Such hotly debated environmental issues as the aggressive extermination of nuisance birds, invisible forms of toxic pollution, the consequences of human technologies, and increasing habitat loss are all expressly ignored by the assumptions that underlie the systems of representation manifest in field-guide literature. These perspectives presented in field guides have environmental ramifications and political valences.

These multiple assumptions take special work to unpack in part because field guides are hybrid texts, existing at a textual crossroads between descriptive nature writing and the visual and textual rhetoric of science.

Field guides make claims about the natural world through images, text, document design, and other content elements such as range maps, and through the relationships between these various parts. Because of this textual hybridity, and because field guides are positioned as reference materials, they have not typically been thought of as environmental literature. However, I argue that field guides make a particular contribution to what one of the founding contributors to the field of ecocriticism, Lawrence Buell, has referred to as the *environmental imagination.*

In the field of ecocriticism, environmental literature is seen as important in shaping larger frameworks of environmental philosophy. Buell's influential 1995 book, *The Environmental Imagination: Thoreau, Nature Writing, and the Formation of American Culture,* helped initiate a wide range of work on environmental literature. Buell argues for seeing authors such as Henry David Thoreau as integral to the formation of a particularly American strain of environmental thought. A good deal of work on environmental literature has extended Buell's interest in literary fiction and nonfiction (Heise 2006, 2008). That reference materials like field guides have not been a larger part of this scholarship is somewhat surprising, however, because in the beginning of *The Environmental Imagination,* Buell calls for the study of a wide variety of texts so as to understand how nature has been theorized and constructed. He writes, "I am ultimately less interested in Thoreau per se than in the American environmental imagination generally, meaning especially literary nonfiction from St. John de Crèvecoeur and William Bartram to the present, but beyond this environmentally directed texts in other genres also" (Buell 1995, 2). Bartram, of course, was an early illustrator of North American birds, creating precursors to the modern field guide.

Patrick Murphy, in *Farther Afield in the Study of Nature-Oriented Literature,* picks up on Buell's suggestion, identifying nature writing and environmental literature as "two different modes of writing about nature and human-nonhuman relationships" that get expressed in fiction and nonfiction alike (P. Murphy 2000, 49). As a result, Murphy argues, genre should not limit the scope of ecocriticism. Buell's and Murphy's calls to move to "texts in other genres" capture how this book, *Binocular Vision,* functions as a study of environmental thought: not through attention to traditional literature but by examining field guides as everyday texts. My explicit argument is for seeing the importance of field guides in reifying and structuring environmental thought, but implicitly I am claiming that the kinds of texts that get studied as environmental literature can be expanded.

Given their strong reliance of visual rhetoric and because they employ scientific systems of representation, field guides have distinct valences within this environmental literature. With the publication of the first field guides in the 1880s and '90s, field-guide authors made connections between birdwatching and systematics, which was, at the time, the main preoccupation of scientists working in the new discipline of ornithology (Barrow 1998).[6] Visually identifying where individual birds fit in a predetermined scientific taxonomy has remained the focus of field guides, even as ornithology has developed into a discipline incorporating many more aspects of scientific inquiry.

In a way, field guides archive and preserve the nineteenth-century ornithological interest in systematics, with one difference being that field guides promote hands-off, largely visual identifications. Ornithology was amenable to the growth of birdwatching as an amateur science in part because, as Yolanda Texera Arnal explains, "Among all the zoological species, birds are the easiest to observe and study. Their diurnal habits, the songs, and visual features make them conspicuous" (2002, 601). But the "science" of ornithology is not as undisputed as representations of birds in field guides make it seem. Species are perpetually being reclassified, merged, and split by the American Ornithologists' Union, and as Alan Gross finds, in a chapter on ornithological discoveries in his book *The Rhetoric of Science*, some science involves processes of what he calls "tendentious simplification" (Gross 1990, 45). In his study of the "discovery" of a species of hummingbird, Gross shows that there can be an all-too-easy collapsing of biological complexity in the effort to advance scientific knowledge about birds. Field guides, by being so closely linked with ornithology, rely in many ways on these forms of tendentious simplification. The scientific leanings of field guides may make them seem authoritative, but that authority is no less constructed or questionable than the literary aspects of the texts.

It is not always the case that when people go outdoors to engage in an environmental hobby or pastime they bring a piece of literary science writing with them. Mountain bikers and kayakers, people who fish or enjoy trail running—these outdoor enthusiasts may read magazines or books about their environmental hobbies, but written materials are not immediately central to the accomplishments of biking down the trail or catching a fish. Like those engaged in orienteering, who constantly rely on maps and global positioning systems to navigate through a challenge, birdwatchers identify new birds with the help of one or more field guides.

To emphasize that field guides are not transparent tools, John Swales describes them as first and foremost *linguistic texts* written in ways that readers necessarily interpret and translate (1995). Michael Lynch and John Law argue, in "Pictures, Texts, and Objects: The Literary Language Game of Bird-watching" (1999),[7] that identifying birds with a field guide amounts to a kind of Wittgensteinian language game, with users matching what they see in the "real world" with various modalities of representation featured in the guide. The kinds of taxonomically minded textual games that field guides are part of heightens their authority, to one extent, positioning them at one vital center of birdwatching. As I explain, these literary, scientific, visual texts suggest much more than the name of the little brown bird bobbing down the beach. Field guides represent the beach the bird is on and the extent to which the ocean is or is not undergoing rapid acidification (Rockström et al. 2009), and they even have implications in terms of the effects birdwatchers have on the lives of birds. For all these reasons, field guides are worth looking at.

This book has two main arcs of arguments. The first concerns the environmental implications of representations. I describe how, at the end of the nineteenth century, early field-guide authors adamantly argued for bird conservation; I then move into the twentieth century, when a range of environmental problems facing birds and the planet at large took a backseat to technical representations of birds. This first argument ends with an exploration of the environmental implications of some of the new modalities of representation in electronic field guides. The second arc involves a history of the field guide itself, explaining differences in terms of field-guide content, rhetoric, and design. This argument also begins in the late nineteenth century, when field-guide authors relied on extended narrative descriptions; it moves into the twentieth century, when technical field guides emerged as quick-reference texts; and it ends in the twenty-first century, with the development of electronic field guides. These two trajectories weave themselves into the book's five chapters as follows.

Chapter 1 describes how two prolific early field-guide authors, Florence Merriam and Mabel Osgood Wright, promoted birdwatching as a set of practices and sensibilities intended to help save birds. In the 1880s and '90s, Merriam and Wright represented the new pastime of birdwatching as a blend of science, emotion, and a deep concern for social issues. A variety of forms of anthropomorphism, in particular, helped these authors make birds seem newly valuable on human terms. Field guides in these early years were certainly not about loving all birds, though; just the "good" ones were favored, and proper citizenship became a powerful lens through which birds were understood. Such anthropomorphic ways of thinking about birds in early American field guides are consistent with what continue to be popular ways of justifying bird conservation based on human benefit. In addition, Merriam and Wright show how, right from the start, saving some species of birds was linked to the purposeful and aggressive management of others.

But field guides and birdwatching soon changed. In chapter 2, I describe how, by the 1930s, field guides became increasingly technical and visual, dispensing with long narratives and turning to images. These new representations of birds and nature constructed the birdwatcher's object of study as separate from cultural, legal, and environmental issues affecting birds. The new field guides, marked by the publication of Roger Tory Peterson's *A Field Guide to the Birds* in 1934, were books to be quickly paged through and scanned. Although they seem unbiased and objectively scientific, I argue that technical field guides are full of assumptions about science, nature, and the environment. Technical field guides silently consent, for instance, to the ongoing management of what continue to be perceived of as "good" and "bad" birds. To show how such assumptions play out, I consider various competing representations of the Bald Eagle, Mute Swan, gulls, and crows. I look at these birds through field guides, environmental law, and popular writing to show how long-standing biases about "good" and "bad" birds continue to inform the aggressive and sometimes arbitrary management of nuisance birds. Binocular vision has developed in ways that encourage birdwatchers to ignore the politics of avian management while profiting from the aggressive culling of birds constructed as a nuisance.

Whereas the technical field guides discussed in chapter 2 picture birds in undisturbed, hygienic natural scenes, chapter 3 explores the varied environmental politics of featuring birds in altered landscapes. I turn to three sources for representations of this kind. The first is Jack Griggs's environmentalist field guide, *All the Birds of North America* (1997), a radical work featuring some species living alongside technology, trash, and humans. In Griggs's unusual field guide, identifying birds is portrayed as part of larger and more involved forms of environmental study that even require attending to the environmental impact of birdwatchers. Included in the second set of images I discuss is a series of General Electric ads featuring altered paintings of birds by John James Audubon. The GE images portray birds living peacefully alongside such environmental hazards as jet aircraft and smokestacks, thus attempting to position GE as a green, bird-friendly corporation. As a point of comparison to GE's message, I turn to the work of contemporary painter Alexis Rockman, whose technical representations of birds depict disastrous encounters between birds and humans that, in some cases, render birds unrecognizable. Chapter 3, then, is an exploration of how technical representations of birds have been used not only to promote birdwatching, but to sustain a range of environmental messages.

In recent years, field guides have jumped from print into a range of electronic technologies. There are now handheld electronic birdwatching gadgets and a range of online field guides, many of which are interactive and networked. This jump to e-guides has presented users with many new and revitalized approaches to birdwatching. In addition, the jump to e-guides creates an opportunity to recognize in new ways the long-standing importance of technology of the book to birdwatching. As chapter 4 shows, book technology has built into it certain affordances and limitations, and they have helped construct birdwatching and binocular vision as particularly visual. To explain this relationship, this chapter focuses on the new availability of recorded bird vocalizations in e-guides that encourage birding by ear in addition to birding by eye. Bird vocalizations, although highly revealing about most species, have been deemphasized in print guides, given the challenges of presenting them on the printed page. So although e-guides present users with new ways of searching for and thinking about birds, they also reveal the extent to which the highly visual disposition of binocular vision has developed in close relationship to the gadgetry of environmental interaction.

Chapter 5 takes its inspiration from one of the more unusual field guides I began this introduction with, William Tice's *A Birder's Guide to the Sewage Ponds of Oregon: Creatures from the Brown Lagoons* (1999). The chapter is a study of textual representations of competitive birding at toxic, polluted sites. I pair reports of birding at three competitive birding events with three toxic sites: big-day birding at EPA Superfund sites, big-year birding at industrial sewage treatment facilities, and listing at toxic landfills. Although birdwatching generally relies on and is thought to contribute to environmental protection, and although competitive birding is frequently connected to green fund-raising events, I argue that representations of competitive birding at these kinds of toxic sites discount the seriousness of toxic pollution as an environmental problem. This final chapter, then, is an exploration of how binocular vision can, particularly in the context of competitive birding, be connected to practices that run counter to conservation. As competitive birders seek more and more birds regardless of where those birds are, substantive forms of environmental intervention are represented as irrelevant. As I describe, serious toxins exist in the lands where many birds live and pass through, but competitive birding practices can function as a text that diminishes the importance of these problems.

To explore, on a small scale, some of the ways that birdwatchers use their field guides, I conducted a qualitative study of birdwatchers at one such toxic site, the Montlake Landfill in Seattle. I then augmented those findings with an online questionnaire responded to by 225 respondents. I conclude the book with a discussion of these findings, focusing on the ways that many of the birdwatchers I contacted alter and modify their field guides. Among the birdwatchers I studied, I found a good deal of variation and play in their textual practices of birdwatching. I take this modification as an optimistic metaphor for the variable ways that field-guide literature can be taken up and responded to as flexible, inhabitable texts.

Taken at face value, field guides are intended to help birdwatchers identify and learn more about birds, but representations of birds and the environment in field-guide literature have much broader implications. In this historical and multidimensional study of field guides, I make the case that everyday texts like the field guide should be taken seriously as interpretive artifacts. This belief derives from my claim that field guides contribute to a much broader environmental imagination. In the case of field guides, the enchanting visual texts we birdwatchers know so well are filled with assumptions about birds and nature that are well worth exploring if we are to understand the importance of these reference materials in their politics of environmental representation.

A BRIEF NOTE ON SCOPE AND TERMS

There are many kinds of field guides. Some cater to beginning birdwatchers and others to advanced birders; terse guides about backyard birds are available, as are voluminous guides to the birds of an entire continent; there are field guides to the birds of single countries (Costa Rica, for instance) and others spanning many nations, such as guides to the birds of Great Britain and northern Europe. This book is a study of field guides dealing with what John James Audubon termed "the birds of America." Audubon meant North America, but it is worth noting that, in part because birds seldom stay in one place, even North America is not as well defined in field guides as one might expect. Species found exclusively in Mexico and Central America are typically excluded from guides to North American birds, whereas other birds (such as the Wood Warbler) that spend most of their time in Asia or Africa but occasionally wind up in Alaska are sometimes included.

Although the focus of this book is on field guides, at times the discussion moves to the textual mediation of the more sociological practices of birdwatching. My use of the term *birdwatching*, which exists throughout the book in the interest of simplicity, is intended to capture, not efface, the many differences among birdwatchers. The term *birdwatching* generally refers to the widest range of bird-identification activities—everything from backyard birdwatching to competitive listing—but there are other, more specific terms. *Birding* is generally used to describe a smaller subset of people who do such things as keep detailed lists of the birds they identify and sometimes participate in organized birding events. There are also *twitchers*, a term used largely in Great Britain, who actively search for rare birds (Moss 2004, 263–95). Birdwatching can be a hobby, an environmental pastime, a form of amateur science, or some combination of these pursuits. There are many psychological, spiritual, emotional, scientific, and environmental motivations for birdwatching, all of which are outside of the scope of this book. But it is worth noting that some birdwatchers watch birds for spiritual and personal reasons, others because birds are seen as aesthetically appealing, others as part of birding competitions, and still others to contribute data to citizen science birdwatching projects that collect quantitative and qualitative statistics about birds for researchers working on population studies (Maynard 2009).

CHAPTER ONE

Field Guides and the New Hobby of Birdwatching

In the 1880s and '90s, someone interested in reading about the birds of New England would have been able to cobble together an array of resources and learn a good deal. There were two comprehensive ornithological manuals in print at the time: Elliot Coues's *Key to North American Birds* (1872) and Spencer Baird, Thomas Brewer, and Robert Ridgway's *History of North American Birds* (1874). Both feature detailed information about bird anatomy and distribution. If one were looking for more experiential narratives about birds, nineteenth-century writers had produced many volumes of natural history writing about birds. William L. Baily's *Our Own Birds: A Familiar Natural History of the Birds of the United States* (1869) and John Burroughs's popular *Wake Robin* (1871) are examples of this writing, focusing on the beauty of birds and creating a sense that communing with birds and nature is transcendental.[1]

Early works of bird illustration could be helpful as well, including expansive projects such as Mark Catesby's *The Natural History of Carolina, Florida and the Bahama Islands* (1731) and John James Audubon's *The Birds of America* (1840). Nineteenth-century periodicals included brief sketches about birds, with *Arthur's Home Magazine* running a series called "Chapters on Birds" (1854, 88–89). Ambitious readers might even discover that there was a good deal of information about game birds in field guides for hunters such as Gurdon Trumbull's *Names and Portraits of Birds Which Interest Gunners, with Descriptions in Language Understanded of the People* (1888). Trumbull's hunting guide progresses through a list of common game birds species by species, describing each one with brief, vivid, classifying language.

Although these resources could be pieced together, compared, and used to help identify birds, none of them could accurately be called a birdwatching field guide. In the 1880s, there was not yet a single resource about the birds of North America that was for nonspecialists, covered all or most species in the region, was portable and easy to use, and focused on identification by sight and/or sound. In the 1880s and '90s, in the context of what appeared to be the impending extinction of dozens of bird species, several authors filled this niche by creating a new genre of the birdwatching field guide.

These new field guides recontextualized and reframed existing information about birds, presenting it in new ways and for new purposes. Not all the authors of these new birdwatching field guides were equally preoccupied with saving birds, and avian protection was closely connected with lethal forms of avian management. But several new field guides were carefully crafted to get readers interested in birdwatching so that they would become advocates for favored and beleaguered birds. In this respect, the new genre was created to put in motion a chain reaction in which a new kind of book (the genre of the field guide) sponsored new practices (birdwatching) that then altered attitudes intended to reverse what "many knowledgeable observers believed" was a rapid decline in bird populations "to the point that irreversible damage would soon occur" (Dorsey 1998, 165).

In scholarship about how countercultures are promoted and social change is initiated, considerable attention has been paid to social movements (Morris and Browne 2001) and the vital role of public address in making things happen (Stewart 1997). Understandably, when we think of radical social movements, and environmental movements in particular, we tend to think of great speeches, marches, protests, and other forms of direct action. Work on the rhetoric of the environmentalism has focused on the tactics of street protests (DeLuca and Peeples 2002), direct action (DeLuca 1999), and more recently "toxic tours" that work to educate participants about environmental pollution (Pezzullo 2003, 2007).

The story of how field guides helped sponsor bird conservation, however, reveals that not all environmental activism is so overt; part of the conservation movement to protect wild birds at the end of the nineteenth century was quiet, subtle, and based on creating new practices and feelings relating to birds. One of the field-guide authors discussed in this chapter, Mabel Osgood Wright, has received a fair amount of scholarly attention, but most of that work has been on Wright's development of a bird sanctuary, contributions to the Audubon Society, and work as an author of children's literature (Forbes and Jermier 2002). Her work as a field-guide author who promoted birdwatching was environmentally significant as well.

This chapter focuses on two of the first and most popular birdwatching field guides, Florence Merriam's *Birds through an Opera Glass* (1889) and Mabel Osgood Wright's *Birdcraft* (1895).[2] Mark Barrow, in his history of American ornithology, calls Merriam's *Birds through an Opera Glass* "the first of this new generation of field guides" (1998, 156), and Raymond Korpi refers to the guide as "the modern field guide in earliest form" (1999, 14).[3] Wright's *Birdcraft*, published six years after Merriam's guide, was reprinted nine times and remained in print for thirty years. These early guides are scientific/narrative hybrids concerned with promoting the study of birds, and I argue that they were instrumental in promoting attitudes and practices that were and have remained key to birdwatching and environmental conservation.

In no way did field guides and birdwatching "save birds" single-handedly, and I do not mean to diminish the importance of bird-preservation groups (such as the National Audubon Society), bird sanctuaries (such as Pelican Island, Florida, established in 1903), and protective legislation (such as the Weeks-McLean Act of 1913 and the Migratory Bird Treaty Act of 1918). Where field guides have been significant is in promoting attitudes, tastes, customs, and, most important, *practices* critical to conservationism and environmental thought. These attitudes, tastes, customs, and practices began to inform and create what I am calling binocular vision, a way of thinking about and relating to birds in particular and "nature" on the whole. As this chapter shows, that larger mindset of binocular vision is one that was set forth by early field-guide authors in ways that were sometimes problematic although strategically chosen to foster a broad acceptance of new forms of environmental conservation.

It may seem strange to think of a genre such as field guides and a hobby such as birdwatching as influencing environmental law, but as Kurkpatrick Dorsey describes in his detailed history of the first major piece of federal legislation protecting North American birds, the Migratory Bird Treaty Act of 1918, the treaty "was a child of sentiment, in that it grew out of the desire to save birds. It was not a response to an international economic disagreement, as [other] treaties had been" (Dorsey 1998, 165). Coordinating with other organized efforts to save birds at the end of the nineteenth century, field guides were instrumental in sustaining ways of thinking about and relating to the environment that played a part in altering social, economic, and environmental landscapes. The new genre of the field guide sponsored an array of new dispositions, practices, and tastes—all contributing to binocular vision—and as a result, field guides helped demonize several forms of overhunting that had imperiled dozens of species (Dorsey 1998, 170–71).

At the same time, field guides as they first promoted birdwatching were not about loving and appreciating all birds; these texts modeled a way of thinking in which the aggressive management of "bad birds" was important. The place of nuisance birds in representations of birds, environmental policy, and our larger environmental imagination is one developed in this chapter and expanded in the next. Although the new genre of the field guide does not, on the face of it, always appear overtly political, and although most contemporary birdwatchers might not think of field guides as conservationist texts, the early field guides I discuss in this chapter were radical in attempting to sponsor new forms of environmental conservation. In some ways, these books asserted their power covertly by seeming to be "only" interested in creating a new hobby, but it is an undercurrent of this chapter and an argument of this book that environmental hobbies have consequences.

To succeed in getting birdwatching off the ground, popular and familiar forms of anthropomorphism, sentimentalism, taxonomic discourse, nationalism, racism, and taste were all mobilized within the new genre to make watching birds seem worthwhile. Merriam's and Wright's guides drew their reader's attention precisely and obsessively to birds, and both books exemplify how early field guides functioned tactically, working to infect a readership with an infatuation for birds that would transform consumers of birds and bird feathers into their caretakers.

It is certainly not the case that birdwatching emerged all at once with the publication of the first field guides to birds in the 1880s and '90s. Instead, it is more accurate to think of bird study as having many long-standing traditions and of *birdwatching*, the popular pastime, having taken shape when it did as a result of four factors: daunting threats to many birds, the emergence of the new scientific discipline of ornithology, a new acceptability for hobbies and leisure-time pursuits in American culture and daily life, and the textual traditions of natural history writing and scientific bird illustration. The threats to birds and features of natural history writing are discussed in subsequent sections on Merriam and Wright; here I sketch some backgrounds for the origin of birdwatching in terms of the formation of ornithology, changes in ideologies associated with hobbies, and conventions of scientific bird illustration.

Jeffrey Karnicky writes that "approximately at the turn of the nineteenth century, the visual encounter between humans and birds, which has been going on since both forms of life have existed, began to solidify into a hobby, into something that a middle-class citizen of America might spend a morning doing" (2004, para. 1). A key reason birdwatching took shape as a hobby when it did was because watching and identifying birds was gaining new scientific interest with the creation of the new discipline of ornithology. Although the scientific study of North American birds dates to the eighteenth century in North America, with the publication of Catesby's *The Natural History of Carolina, Florida and the Bahama Islands* (1731), and although ornithology had taken shape in Europe by 1850 (Farber 1982/1997), North American ornithologists founded the American Ornithologists' Union (AOU) in 1883 (Barrow 1998; Battalio 1998, 36). Soon after, the group's journal, *The Auk*, was first published. The early members of the AOU were mainly concerned with systematics—finding, sorting, and naming species and subspecies—and they studied bird specimens more than they did living birds.

By being linked to the science of ornithology via the common concern of species identification, birdwatching gained a certain formal status and legitimacy for itself while giving participants an accessibly challenging set of tasks to accomplish. Birdwatching established itself as a form of amateur science to the extent that it involved the careful practice of assigning individual birds to taxonomic categories. And as citizen scientists, who can be defined as nonscientists who make contributions to scientific research, many birdwatchers took part in the popular Christmas Bird Count (still ongoing) to contribute data to longitudinal distribution studies (Maynard 2009).

Birdwatching also emerged when it did because hobbies were gaining a new, approved status in American culture. In his cultural history of hobbies, *Hobbies: Leisure and the Culture of Work in America* (1999), Steven Gelber advances a Marxist explanation of the process through which hobbies became newly legitimated in the United States during the period 1880–1910, precisely the time when birdwatching was taking shape as a hobby. "Before about 1880 a hobby was a dangerous obsession," Gelber writes. "After that date it became a productive use of free time" (3). Gelber argues that hobbies gained acceptability when they did not because they questioned the central values of society, but because hobbies affirmed them by amounting to a form of "productive leisure" in maintenance to the larger system of capitalism.[4]

Gelber's argument applies in part to birdwatching, because identifying birds in the out-of-doors means adopting a task-minded and methodical orientation toward the natural world. When birders diligently identify birds and keep lists, they can be said to engage in a mode of accountancy that puts birdwatching in sync with the task-oriented workplace and well-managed home. A logical extension would be the conclusion that, as an environmental practice, birdwatching endorses the larger capitalist and societal status quo that, in many cases, impinges on some of the natural resources that are important to birds. But, as I go on to discuss, early formulations of birdwatching were also an affront to aspects of commodity culture that resulted in the deaths of many birds. Furthermore, the spiritual and aesthetic connections many birdwatchers have established with birds run counter to a strictly capitalist valuation of nature based on use value.

Birdwatching could not exist in the ways it does without the field guides at the center of the pastime, and, like birdwatching itself, field guides did not appear out of nowhere in the 1880s and '90s. Instead, the first field-guide authors drew on established textual traditions to make a new kind of book that borrowed from several existing textual traditions. Natural history writing was combined with scientific bird illustration in the development of the first field guides. Such bird illustration is an artistic and scientific tradition that had been integral to the cataloging of North American flora and fauna since the sixteenth century.

In fact, on first glance Mark Catesby's *Natural History of Carolina, Florida and the Bahama Islands* (1731) looks a lot like a field guide. Catesby's book takes a naturalist's perspective to describe regional flora, fauna, land features, and people in colonial America, including detailed color plates and classificatory descriptions of many bird species (Catesby and Feduccia 1985, 11). Whereas field guides are quick-reference books, meant to be flipped through and used for quick identifications, books like Catesby's are comprehensive. Catesby's illustrations were simple, sometimes unlifelike, and presented individual named species in poses that a reader could use to become acquainted with primary characteristics. Catesby's *Natural History of Carolina, Florida and the Bahama Islands* includes early forms of taxonomic discourse, both in the alphabetic descriptions and visual images. By removing species from their environmental contexts and displaying and describing unique diagnostic features, the descriptions and images work to mark the individual as a representative of a species with a place in a taxonomy.

Although late-nineteenth-century field guides are largely narrative and thus informed more by works of natural history than the writing in more classificatory texts such as Catesby's, the scientific language in *The Natural History of Carolina, Florida and the Bahama Islands,* which is characteristic of formal and scientific bird study, still influences the text in many field guides. The writing is terse, is full of descriptive features that are often comparative, focuses largely on what can be seen, and features a specialized lexicon. Catesby describes "The Parrot of Carolina" (what became known as the Carolina Parakeet, now extinct) in this way:

> This Bird is of the bigness, or rather less than a Black-bird, weighing three ounces and a half: the fore-part of the Head Orange-colour; the hind part of the Head and Neck yellow. All the rest of the Bird appears green; but upon nearer scrutiny the interior vanes of most of the wing-feathers are dark-brown. (Catesby 1731, 16)

Describing the Myrtle Warbler 170 years later in her 1895 field guide, *Birdcraft,* Mabel Osgood Wright echoes Catesby's taxonomic discourse, although Wright's phrasing is even more streamlined (having fewer verbs):

> Slate colour, striped and streaked with black. Crown, sides of breast, and rump yellow. Below whitish; upper breast black. Two white cross-bars on wings; tail with white spot. In winter, brownish olive; yellow of rump constant, but lacking on crown and breasted. Bill and feet black. (Wright 1895, 96)

I refer to this writing as taxonomic because it is actively engaged in the systematic classification of organisms. Although neither description amounts to a complete scientific taxonomy of the Carolina Parakeet or the Myrtle Warbler, such descriptions provide the elements needed to sort and name a bird in the field by sight. Writing of this kind is one clear antecedent that field-guide authors draw on and reanimate in working to promote birdwatching as a practice of precision.

Such precision is similarly found in Alexander Wilson's *American Ornithology* (published in nine volumes between 1808 and 1814; republished with additions by Thomas Mayo Brewer in 1840) and even later ornithological manuals such as Elliot Coues's *Key to North American Birds* (1872) and Baird, Brewer, and Ridgway's *History of North American Birds* (1874). Wilson's *American Ornithology* spans 746 pages in its 1840 edition and is a dense manual made up of full-page plates comparing many similar-looking species. Wilson's book also provides many taxonomic classificatory details, includes natural history descriptions, and reproduces extensive quotations from other authors writing about birds. This kind of comprehensive scientific material informed field guides but did not end up overdetermining them to the point that field guides became unusable as portable, quick-reference texts.

Standard visual practices for field-guide images came from Catesby and Wilson's illustrations, but even more so from John James Audubon's bird illustration and natural history, first published as *The Birds of America* (1840). Although Audubon's volume was by no means a portable text meant for those interested in identifying birds in the field, his unique approach to realistically posing birds included staging bird portraits in "natural" settings, separating birds by species in each image, and painting birds in poses that prominently displayed unique and revealing characteristics. Audubon's images are descriptive in two ways: they tell a story, and they identify each species of bird. Many late-nineteenth-century field-guide images continued to bridge between these two not necessarily related goals—satisfying, that is, the expectations of narrative natural history and taxonomy—in combining the seriousness and validity of the scientific discipline of ornithology with the compassion engendered through sentimental description.

Birdwatching emerged when it did because of a range of factors that inflected the conservationist leanings and dispositions of the pastime. The new science of ornithology, for instance, gave birdwatching a technical bent; the new acceptability of productive leisure meant that birdwatching developed habits of mind and practice that were in dialogue with larger cultural systems and values; and the textual traditions of natural history writing and scientific bird illustration meant that birdwatching tapped into preexisting systems of visual and narrative representation. Florence Merriam and Mabel Osgood Wright found ways to inflect this matrix of origins that allowed them to represent birdwatching as something fun to do and as an environmental intervention.

FLORENCE MERRIAM AND MABEL OSGOOD WRIGHT

Florence Merriam and Mabel Osgood Wright were two extraordinarily prolific writers. Florence Merriam (1863–1948) had a long career as an ornithologist, naturalist, environmental activist, and writer, publishing more than ten books and numerous articles beginning in the 1880s and extending into the 1940s.[5] Her books on birds include what is often referred to as the first field guide to North American birds, *Birds through an Opera Glass* (1889); a personal narrative about bird study in the west, *A-Birding on a Bronco* (1896); a work of natural history, *Birds of Village and Field* (1898); and two regional bird guides, *Handbook of Birds of the Western United States* (1902) and *Birds of New Mexico* (1928).

Merriam began writing about birds in her early twenties (ten years before marrying the accomplished naturalist and mammalogist Vernon Bailey, in 1899, and changing her name to Florence Merriam Bailey) and demonstrated a commitment to writing about the roles of women in society at a time of several pressing reform movements. In an era when many women writers adopted pseudonyms (Kofalk 1989, 33–36), Merriam did not. That is no small point, as in many ways Merriam's writings about birds align the plight of female birds with that of women. *Birds through an Opera Glass* (1889), which was Merriam's first book, deals with seventy of the two hundred to three hundred species commonly seen in the eastern United States. Merriam began the work that would become the guide while she was a student at Smith College, publishing a series of articles titled "Hints for Audubon Workers: Fifty Birds and How to Know Them." These pieces ran in George Grinnell's short-lived *Audubon Magazine*, a publication of the first Audubon Society and precursor to what would become the long-running *Bird-Lore* (established in 1899). Writing about birds was just a part of Merriam's environmental activism, though: at Smith she led bird walks and was an avid letter writer on the topic of bird preservation (discussed at length in Kofalk).

Mabel Osgood Wright's popular field guide was the third of her more than thirty books. *Birdcraft* (1895), which bears the long subtitle *A Field Book of Two Hundred Song, Game, and Water Birds, with Full-Page Plates Containing 128 Birds in the Natural Colours, and Other Illustrations*, was released six years after *Birds through an Opera Glass* and thus came out during a period of expanding interest in birdwatching. A more widely used guide, *Birdcraft* was ultimately reprinted nine times before, in the 1930s, the popularity of Roger Tory Peterson's technical, visual field guide (discussed in chapter 2) pushed the narrative-based *Birdcraft* out of print (Gibbons and Strom 1988, 183).

Birdcraft, though, was only one of Wright's many projects relating to birds and nature. She wrote a regular column for and coedited the National Association of Audubon Society's *Bird-Lore* (Gibbons and Strom, 1988, 182), published a guide to flowers and ferns (Wright 1901a), published a book of nature writing (Wright 1894), wrote about other animals in *Four-footed Americans and Their Kin* (Wright, Chapman, and Seton 1898), and collaborated with ornithologist Elliot Coues, author of the field manual *Key to North American Birds* (1872), to produce *Citizen Bird* (Wright, Coues, and Fuertes 1897), a book for children with anthropomorphized birds as main characters. What gained Wright the most renown, however, was her domestic fiction (Wright 1897, 1901b, 1903, 1904, 1910). Near the end of Wright's life, she published an account of the founding of her Birdcraft Sanctuary (1922) and her autobiography, *My New York* (1926).

Most of Wright's books are now out of print, but her Birdcraft Sanctuary (established in 1914 and taking its name from her popular field guide) still exists as an early example of private land stewardship as an approach to wildlife conservation (Gibbons and Strom 1988). As Daniel Philippon makes clear in his chapter on Wright in *Conserving Words: How American Nature Writers Shaped the Environmental Movement*, "Both men and women, Wright understood, were more likely to protect endangered places and creatures if they could first learn to recognize the fundamental connections between ecological and social communities, a belief well illustrated by the title of Wright's 1906 book *The Garden, You, and I*" (Philippon 2004, 89).

Merriam's publishers positioned the first field guide, *Birds through an Opera Glass*, as an educational book for maturing and mature readers.[6] In its two editions, which were virtually unchanged, Merriam's guide remained slim and pocket-sized (seven and a half inches by four and a half inches by one inch), bound in a simple hard binding with blank endpapers. The guide consists mainly of text with the occasional black-and-white wood engraving set in with the type (figure 1.1). Because of the dearth of images, small size, and simple binding, the book would have been fairly inexpensive. Because the purpose of Merriam's guide was to reach the widest audience possible, allowing for maximum social impact, these practical features made sense. Subsequent guides included more images, but the reliance on narrative in this first guide allowed for the kinds of complex rhetorical appeals Merriam makes. *Birds through an Opera Glass* has three main sections: a brief introduction, seventy species descriptions (each one to five pages in length), and several appendixes. The introduction functions mainly to sell birdwatching, and Merriam's pitch for the new pastime focuses on the value and importance of classification. "The first law of field work is *exact observation*," she writes, because "you will find it much easier to identify the birds from your notes than from memory" (Merriam 1889, 3). Merriam suggests that the birdwatcher carry both opera glass and notebook into the field, tools that mark a competent birdwatcher and define birdwatching as anything but frivolous.

Although these two field guides, *Birds through an Opera Glass* and *Birdcraft*, share many features, Wright distanced herself somewhat from Merriam's earlier guide by setting out a slightly more comprehensive and scientific set of descriptions of birds.[7] In this way, Wright's approach to birdwatching leaned more toward amateur science. Wright's guide covers many more birds than Merriam's, and each species description

70 *BIRDS THROUGH AN OPERA-GLASS.*

crest make him look! There! he is off again, and before we think where he is going we hear the echo of his rousing *phe-phay, phe-phay* from the depths of the woods.

In many places the jays are common winter residents, pitching their tents with the hens and barnyard animals and comporting themselves with familiar assurance. But in this region they are irregular guests. Sometimes they are here for a few days in the fall, or visit us when the hawks

BLUE JAY. 71

return in spring, teasing the young observer by imitating the cry of the redtailed hawk. But if the fancy takes them they spend the winter with us, showing comparatively little of the timidity they feel in some localities.

Last fall a party of jays stayed here for some time, but when I was congratulating myself on having them for the winter, they left, and did not return till the middle of January. Then one morning one of them appeared suddenly on a tree in front of the window. He seemed to have been there before, for he flew straight down to the corn boxes by the dining-room. The gray squirrels had nibbled out the sweetest part of the kernels, and he acted dissatisfied with what was left, dropping several pieces after he had picked them up. But at last he swallowed a few kernels and then took three or four in his bill at once and flew up in a maple. He must have deposited some of them in a crotch at the body of the tree, for after he had broken one in two under his claw — striking it with "sledge-hammer blows" — he went back to the crotch, picked up something, flew back on the branch, and went through the process over again. The second time he flew down to the corn boxes he did the same thing — ate two or three kernels, and then filled his bill full and flew off — this time out of sight. Since then I have often seen him carry his corn off in the same way, giving his head a little toss to throw the kernels back

Figure I.1. Sample two-page spread including one of the few wood engravings (of the Blue Jay) in Florence Merriam's *Birds through an Opera Glass*, 1899. From the collections of the University of Illinois, Urbana-Champaign.

begins with a brief taxonomic section, focused on individual diagnostic attributes, followed by a longer narrative. Although anthropomorphism is greatly diminished in *Birdcraft*, Wright still describes some birds in human terms and favors species that embody cultural virtues. "Spend a month on the bird-quest, or a week even, and your eyes will be opened to the possibilities," Wright opines in her introduction, "and you will become alive to the fact, that the feathered race has its artisans the same as the human brotherhood" (Wright 1895, 14). As I will discuss, Wright is adamantly in favor of these productive bird-citizens (here "artisans"), which she renders as virtuous birds, and the flip side of this is that she violently condemns birds she sees as in any way diminishing beauty or industriousness.

Birdcraft is a bigger book than *Birds through an Opera Glass* (six inches by eight inches by one and three-eighths inches, at 317 pages). In its first edition, *Birdcraft* included fifteen two-page color inserts (multicolor chromolithography printed on coated paper, tipped in during assembly, and glued to one-quarter-inch stubs). These folding two-page color inserts, painted by Louis Agassiz Fuertes, add somewhat of a visual dimension to this field guide, although it still relies mainly on text.[8] Wright's guide begins with four extensive introductory chapters, loosely based on the annual cycle of seasons ("The Spring Song," "The Building of the Nest," "The Water Birds," and "Birds of Autumn and Winter"), and then offers brief informative sections titled "How to Name the Birds" and "Synopsis of Families." Following this introductory material are Wright's "Bird Biographies," the species-by-species accounts in which Wright describes each species individually. These descriptions are sorted by Wright's preferences, with favorites coming first, in the following order: songbirds; perching songless birds; birds of prey; pigeons, quails, and grouse; shore and marsh birds; and swimming birds. At the back of the volume is an apparatus for identifying birds by color, habitat, and other features, and then two indexes with English and Latin names.

Jennifer Price, in *Flight Maps: Adventures with Nature in Modern America*, describes some of the cultural contexts within which Florence Merriam's and Mabel Osgood Wright's environmental activism can be understood. Price describes efforts to end what was called the plume trade as part of a larger social movement, in the second half of the nineteenth century and early years of the twentieth, involving organized groups of middle-class women agitating for progressive causes and social reform (Price 1999, 2004). The plume trade involved the production and sale of abundantly festooned bird hats. As women's labor historian Wendy Gamber writes, "by the mid-nineteenth century, artisans had largely been replaced by shopkeepers," and this form of capitalist "progress" meant that milliners in major East Coast cities could stock many hundreds of mass-produced hats adorned with the latest fashions in birds and bird feathers (Gamber 1997, 177–82). Bird hats were in fashion, and the high demand resulted in dramatic drops in bird populations.

Both Merriam and Wright responded to the plume trade not with accusations directed at the women who chose to wear bird hats, but with a positive strategy of describing living birds as worth appreciating. Buying bird hats is less directly countered than indirectly undermined through the sponsorship of new activities based on appreciating living birds. This strategy targeted the white, middle-class, East Coast women who set fashion trends by wearing birds on their hats. Merriam's and Wright's guides appealed to such readers on emotional, ethical, and aesthetic grounds. Whereas later field guides became increasingly focused on identification, the first field guides to North American birds served that purpose and more as Merriam and Wright worked to both describe which birds were which while justifying why they thought it worthwhile for consumers to alter their tastes.

Although Merriam and Wright cast birdwatching as an ethical pastime, they and other early field-guide authors were careful also to appeal to the self-interests of birdwatchers. Merriam writes in her introduction to the 1899 edition of *Birds through an Opera Glass*, "The student who goes afield armed with opera-glass and camera will not only add more to our knowledge than he who goes armed with a gun, but will gain for himself" (1889, v). In this passage, Merriam constructs birdwatching as not merely a casual pursuit, but a way for the amateur scientist to "add to our knowledge" of ornithology. In this way, birdwatching could be serious business through being constructed as not only drawing from scientific knowledge but as contributing significant findings to science (Barrow 1998, 165–81).

Birdwatching was also described as emotionally, spiritually, and ethically beneficial to birdwatchers. Although Merriam does not directly accuse the wearers of bird hats of doing wrong, her rhetoric does become pointed when describing careless hunters and reckless boys, the latter having developed a reputation throughout the nineteenth century for what was known as senseless killing—the killing of birds and destruction of bird nests and eggs for the mere pleasure of it.

Imbuing birdwatching with aspects of scientific labor helped disassociate the pastime from more passive forms of leisure, an important move given how, at the end of the nineteenth century, leisure was still seen by many as a threat to the Protestant work ethic (Gleason 1999). Even though Merriam's field guide is not an ornithological manual, her introduction implies that birdwatching should be a form of intensive amateur science, with birdwatchers working on the margins of the newly forming discipline of ornithology. She writes: "Begin with the commonest birds, and train your ears and eyes by pigeon-holing every bird you see and every song you hear. Classify roughly at first,—the finer distinctions will easily be made later" (Merriam 1889, 1). At the time, species classification was the predominate occupation of ornithologists (Barrow 1998; Battalio 1998), and in this way Merriam and Wright both align birdwatching with early ornithology. In one critical way, however, birdwatching was meant to differ from ornithology: the birdwatcher was to identify by eye and ear, whereas the ornithologist collected specimens and identified in part by dissection. Although the scientific discipline of ornithology would, over the course of the twentieth century, come to involve much more than bird identification (Battalio 1998), when birdwatching was first becoming a hobby, Florence Merriam and Mabel Osgood Wright aligned it with the central activity of ornithological science at that point: classification.

In her study of nineteenth-century natural history writing, *The Book of Nature: Natural History in the United States, 1825–1875*, Margaret Welch describes natural history writing in a way that characterizes much of what Wright does in her somewhat more comprehensive and technical *Birdcraft*. Welch writes: "The practice and language of natural history reinforced a way of ordering the world that focused on species, with the individual encountered representing the whole. The characteristics of appearance, locality, and habits that distinguish it from like species receive primary attention. Passages often list physical identifiers such as size in numerical figures, include Latin, and, in the longer prose sections, adopt the third person narrative" (Welch 1998, 166). Most of *Birdcraft* consists of Wright's species descriptions, and each involves the two parts Welch describes: taxonomy and narrative. Entering into what could be called the larger genre system (Russell 1997, 2002) of late-nineteenth-century books about bird identification—including field manuals, ornithological texts, natural history writing, books for children, and field guides—*Birdcraft* is a hybrid text in many ways. Like *Birds through an Opera Glass* and other early examples of the genre, *Birdcraft* borrowed and combined from existing genres in new and not entirely codified ways.

In the discussion that follows, I focus on the ways that *Birds through an Opera Glass* and *Birdcraft* taxonomize birds by pairing those taxonomies with other similarly structured and hierarchical systems such as the society, taste, and emotional register. Both authors combated species destruction by employing an emergent environmentalist rhetoric that positioned readers as emotional saviors of preferred species and enemies of disliked ones.

ANTHROPOMORPHIC BIRDWATCHING

Natural history writing, which often involved anthropomorphism, was a legitimate form of science writing and scientific inquiry in the nineteenth century. Anthropomorphic analogues between birds and humans were integral to legitimizing the study of birds as socially relevant. For Merriam and other early field-guide authors who relied heavily on anthropomorphic descriptions, birdwatching was less about seeing the sublime in nature than it was about interacting with a series of human foils and reflections—a type of looking outward to gaze within. Florence Merriam's use of anthropomorphism connected the study of birds with interests in cultural topics such as fashion, ideal citizenship, and what she broadly terms "woman's wrongs."

Twenty-first-century users of field guides would be shocked to find anthropomorphic descriptions in a contemporary field guide, as anthropomorphizing animals is currently associated with children's literature, fables, and allegories (for example, George Orwell's *Animal Farm*), but it was integral to several early field guides.[9] Florence Merriam uses multiple forms of anthropomorphism in her guide, all of them coexisting alongside and even facilitating the more positivist scientific discourse of classificatory identification. Anthropomorphism is not separate from the science; it is part of it. The different forms of anthropomorphism produce particular emotional and empathic narratives about birds because anthropomorphic justifications for caring about and sympathizing with birds were strategically meant to make killing most species of birds seem unjust for human reasons.

Anthropomorphism was used to help classify birds, and it was also environmentalist in that it was intended to foster bird preservation. Merriam's writing exemplifies how anthropomorphism allowed late-nineteenth-century bird preservationists to incorporate birds into existing systems of ethical sentiment. Merriam's birds were not to be saved because they were an integral part of fragile ecological systems, as later environmental writing would have it, but because killing birds disrupted human spheres of value. Anthropomorphism, evident on nearly every page of Merriam's field guide, is a common feature of other nineteenth-century natural history writing and periodical literature aimed at promoting a variety of reforms from anti–animal cruelty, advanced by the Society for the Prevention of Cruelty to Animals, to abolitionism (Parris 2003).

The varieties of anthropomorphism in *Birds through an Opera Glass* function (as does all anthropomorphism) to render the natural world in human terms. "Mr. Robin," the first species the reader encounters in Merriam's guide, has "city friends" and "is, as everyone knows, a domestic bird, with a marked bias for society. Everything about him bespeaks the self-respecting American citizen" given his "calm, dignified air" (Merriam 1889, 4). The birds Merriam is most fond of—and she does not hesitate to say which ones those are—are described as ideal citizens (referred to with the titles Mr. or Mrs.), a role that works to improve these birds' "reputations" while implying that less ideal citizen-birds (and there are many of these in Merriam's field guide) deserve no protection. In this way, mobilizing citizenship as a corollary between birds and humans enables Merriam to justify the protection of preferred species alongside the extermination of other less preferred ones.

Another respectable citizen is the common and vocal Red-winged Blackbird. Merriam writes, "As Thoreau says, his red wing marks him as effectually as a soldier's epaulets" (Merriam 1889, 91). The anthropomorphism of this simile achieves an equally powerful result as was accomplished with "Mr. Robin": the Red-winged Blackbird is rendered a proper citizen by being likened to a fighting soldier, and by referring to Henry David Thoreau (as Merriam does throughout her text), the citizen-bird is positioned within a tradition of environmental writing. If a common bird such as the Red-winged Blackbird can be likened to a regal soldier, then killing and wearing such a creature on a hat is wrong; engendering pity in this way (and even more important, sympathy) for birds is how this guide and others like it acted to preserve endangered birds. Anthropomorphism has an elevating effect when applied to creatures otherwise seen as insignificant.

Describing plumage as attire—wing bars become epaulets—is a common trope in Merriam's anthropomorphism, and such an equation (feathers equal clothing) serves a dual purpose of helping readers identify birds by sight and aligning birds with fashion-conscious consumers. Merriam's Yellow-rumped Warbler stands out from the other species in the "black zouave jacket he wears" (189), juncos are pious "companies of little gray-robed monks and nuns" (138), and the Bobolink is an even more radical dresser, outfitted in

> shades of short hair and bloomers, what an innovation! How the birds must gossip! Instead of the light-colored shirt and vest and decorous dark coat sanctioned by the Worth of conventional bird circles for centuries, this radical decks himself out in a jet-black shirt and vest, with not so much as a white color to redeem him; . . . But don't berate him—who knows but this unique coloring is due to a process unrecognized by the Parisian Worth, but designated by Mr. Darwin as 'adaptation'? (28)

By citing Charles Darwin (another frequent touchstone for Merriam), Merriam locates her anthropomorphic descriptions within a scientific frame, indicating the way an anthropomorphized scientific mindset is not in the least bit contradictory for her. She also implicates the Bobolink in dress reform, through the mention of bloomers, referencing this flash-point issue of feminist social reform.

By regularly referencing feminist social movements (here, the volatile issue of dress reform), Merriam metonymically makes the preservation of birds into a women's issue, thus attempting to appeal to the predominately female consumers and wearers of bird hats. The "who knows" in the final sentence of this long quotation refers to the birds themselves; unaware of natural selection, they are imagined—with a bit of playfulness—as part of the "Parisian Worth" obsessed only with fads and fashions.

Anthropomorphizing birds in this way, which one could describe as "dressing birds up in human clothing," simultaneously works to make each species distinct and memorable (serving classificatory identification) and worthy of sympathy. A reader can sympathize with these fashionable birds not only because they dress and "gossip" like the rest of us, but because they are identifiable and no longer a confusing or chaotic array; birds that are carefully sorted and named present themselves as species amenable to casual inquiry. Both the anthropomorphism and attention to classification become demystifying in this way, bringing readers closer to both knowable and manageable birds. Furthermore, as a guide aimed at changing taste in the consumer culture of late-nineteenth-century fashion, birds are themselves fashion conscious and thus at least as human as wearers of bird hats. When cast in this light, killing birds (as little fashionable humans) for hats is symbolically misanthropic.

The sartorial anthropomorphism is added to with behavioral anthropomorphism that further portrays wild birds as creatures worth protecting and appreciating in the out-of-doors. Regarding the Least Flycatcher's acrobatics, Merriam writes, "This small bird seems a piquant satire on the days of tournament and joust, when knights started out with leveled lances to give battle to every one they met" (87–88). Like the Red-winged Blackbird who is a soldier, this species is not paired with common folk—workers or the poor—but with regal, idealized citizens. In this way, Merriam promotes birds through the ranks of social importance. Furthermore, in that the flycatcher is a "piquant satire," Merriam acknowledges just how silly such comparisons are while insisting that seeing and being amused by birds in this anthropomorphic way is valuable if one is to create an enjoyable version of birdwatching involving the ability to recognize and name species. Merriam demonstrates a comfort with pairing avian and human behavior, thus creating another way to identify birds while identifying *with* birds. Anthropomorphism of this kind makes the guide entertaining to read, as Merriam jokes not just about the bird world, but about the silliness of our human world by labeling birds with some of the more ridiculous human characteristics.

So far we have seen how Merriam's anthropomorphism works to associate particular bird species with soldiers, knights, and fashion-conscious progressive consumers. Birds are anthropomorphically given idealized citizen status by characterizing them as traditional family members as well. Married bird couples abound in Merriam's text, for instance, ultimately equating bird preservation with the conservation of marriage. Orioles, for instance, are described as newlyweds: "A pair of young and inexperienced orioles fell in love and set out, with the assurance of most brides and grooms, to build a home for themselves." Unfortunately, the "home" literally falls apart, causing Merriam to speculate, "Was it the pressing business of the honeymoon that interfered with the weaving, or was it because this young couple had not yet learned how to pull together?" (54). Jokes about fornicating birds are meant to be humorous, but such projections of human customs, beliefs, and character onto songbirds in particular works to render this group of birds as the most human of all and as caretakers of marriage, fidelity, and love.

If the orioles are faithful lovers, as many of Merriam's birds are, the Kingfisher is a fearless "woodsman" who explores uncharted interiors of the continent. Merriam writes: "Ask him for his compass. He needs no trail. Follow him and he will teach you the secrets of the forest. For here lies the witchcraft of our new world halcyon, rather than in the charming of sailors' lives, or in the stilling of the sea" (60). In this passage, the Kingfisher takes on moral character in excess of being a diving, fish-eating bird; the Kingfisher is an explorer and tamer of the wild. The orioles are married newlyweds and struggling domestics, and the Kingfisher is brave and independent; both are worthy of protection since they amount to, in tandem, an avian fulfillment of domestic and pioneer human virtues. To create new conservationist values, Merriam draws on extant valuations of citizenship and the family.

Birds through an Opera Glass is a guide to much more than birdwatching: it is also a guide to culture. Merriam anthropomorphizes birds so as to put forth social commentary and critique. This technique reveals the broad range of aims Merriam sees a field guide as capable of addressing. Whereas the forms of anthropomorphism discussed so far render birds as human and imbue them with virtue, this related form of anthropomorphism only elliptically comments on birds or birdwatching while focusing much more on human society. There is a jocular quality in this kind of anthropomorphism, linking such comments with Merriam's other good-humored appeals to enjoy living birds.

About the female Black-throated Blue Warbler, a species marked by dramatic sexual dimorphism that can be vexing to birdwatchers—males are a dark bluish black, females are a much lighter yellowish green—Merriam writes:

> Like other ladies, the little feathered brides have to bear their husbands' names, however inappropriate. What injustice! Here an innocent creature with an olive-green back and yellowish breast has to go about all her days known as the black-throated blue warbler, just because that happens to describe the dress of her spouse! . . . Talk about woman's wrongs! And the poor little things cannot even apply to the legislature for a change of name! (187)

The passage uses birds to critique the androcentric naming practices associated with the social institution of marriage. It does so through an informative reflection on how many bird species have been named based on the physical characteristics of more colorful males. Coming at a time when the male-dominated discipline of American ornithology was newly forming and limited women's roles were being publicly questioned and contested, this description of "injustice!" cuts several ways. By commenting on "woman's wrongs"—although safely, in the context of species names—Merriam make birds relevant to the audience of progressive-minded women who would have worn bird hats. In the passage, the plight of female birds misnamed by male ornithologists becomes the plight of women.

A more moderate and mocking undertone in Merriam's description, however, somewhat diminishes or possibly disguises its critique. When Merriam remarks "Talk about woman's wrongs!" in the context of the relative trifle of how birds are named, she can be seen to be taking feminist protests lightly. In this sense, the passage's ambivalence allows it to read both as a commentary on "woman's wrongs" and a parody of one feature of feminist advocacy during the period. It is important to recognize that within a field guide, a genre that in subsequent decades became almost exclusively concerned with aiding in bird identification, Merriam includes social commentary about gender inequality, naming, and power.

As one of the first authors working in the genre of the birdwatching field guide, Merriam treats it as not yet codified around the sole goal of identification, and she therefore renders field guides as flexibly capable of social commentary as well. Merriam's guide does a lot more than merely work to facilitate field identification, and by doing so she uses social commentary to appeal broadly to readers to ultimately represent birdwatching, a new pastime for many, as desirable.

Anthropomorphism that functions as a form of social commentary in Merriam's guide pertains mainly to issues that would have been especially pertinent to many female readers: gender roles, marriage, and raising children. Again, these issues are social categories that render birds as approximating the ideal forms of citizenship that Merriam sees as worthy of protection in and of themselves. Although Merriam is critical of certain aspects of gendered society, she reifies others:

> I am afraid Mr. Goldfinch is not a good disciplinarian, for his babies follow him around fluttering their wings, opening their mouths, and crying *tweet-ee, tweet-ee, tweet-ee, tweet-ee,* with an insistence that suggests lax family government. Someone should provide him with a bundle of timothy stalks! And yet who would have our fairy use the rod? (78–79)

Merriam characterizes the male Goldfinch as a species that takes part in raising the young (such knowledge would aid in field identification), but once anthropomorphized she also critiques his parenting. Merriam seems to endorse corporal punishment, but following that endorsement she critiques the very same practice as not fit for a "fairy." She even goes on to say that "Mr. Goldfinch" has a "secret" better than the whipping with a birch branch. If there are better ways to discipline children than whipping them, Merriam seems to indirectly endorse such methods. Again, she is ambivalent while finding a witty way to make classificatory identification interesting and newly compelling to her readership through social commentary.

In another passage, Merriam recalls a time when she and her classmates at Smith College were surprised to see a male bird tending a nest: "We could not help laughing at this domestic turn, he looked so out of place; but we liked him all the better for minding the babies while his wife took a rest" (154). Although seeing a male bird engaged in what the anthropomorphic lens considers an inversion of gendered labor practices makes the group laugh, they see it as a fitting form of assistance for the "wife." Like her critique of the institution of marriage, here Merriam's social commentary is cautiously moderate. As with the previous examples, interpreting this passage in only one way is a challenge. Does *Birds through an Opera Glass* support the notion of separate gendered spheres or criticize such hierarchies of difference? Is Merriam for corporal punishment or against it? Does she mean to critique the practice of taking the husband's name or the legal action to change names back?

In these passages we have just looked at, the social institutions of naming, marriage, and corporal punishment are incorporated into discussions of avian definitions through uses of anthropomorphism, even if Merriam does not polemically inveigh for or against the social institutions she describes. Birds are known through the lens of feminist social reform (among other things). Defining birds connects in the text to defining social roles and categories, occasioning critique, reinforcement, and commentary on the social through taxonomic encounters with birds. Strategically, the guide appeals to a wide audience to alter conceptions about the value of birds. If the text provided more extreme forms of social commentary, it would have surely alienated Merriam's more conservative readers, defeating her central purpose of convincing all owners of bird hats to put them away, stop buying more, and come to appreciate birds in the out-of-doors. Because of these goals, Merriam's field guide connects birds to the social worlds her readers lived in, reflecting on social institutions while teaching readers how to identify and classify.

As we have seen, Merriam's use of anthropomorphism creates relationships with and appeals to fashion, ideal citizenship, and what she broadly terms "woman's wrongs." Merriam's descriptions also position birds in or in relation to nineteenth-century domesticity and what can be called the happy home. About one of her favorite species, the Chip-Bird (our Chipping Sparrow), Merriam describes how an offense against the happy home results in motherhood going awry: "As the babies grew older I suspect their mothers poisoned [the chicks'] minds, too, for as nearly as I could make out a coldness grew up between the families of infants" (64–65). This description contrasts with many others characterized as "happy homes" held together by a strong female figure. In this way, birds are homemaker-citizens worth enfranchising and protecting. The Chestnut-sided Warbler, for instance, is "altogether sensible, straightforward, industrious, and confiding—a captivating, motherly body" (191).

In reference to Indigo Birds (our Indigo Bunting), Merriam describes how a male Indigo Bird created a "threat of conjugal authority [that] subdued [the female Indigo Bird], and at last she meekly flew off into the woods with him. But, like some other good wives, she had her way in the end" (122). Because the dominated female bird asserts power "in the end," Merriam manages both to describe male/female species dynamics in human terms and to comment that "some other good [human] wives" deploy equally successful tactics in creating happy homes. Birds as mothers and domestic managers are worthy of affection and admiration as Merriam appeals (through anthropomorphic characterization) to her audience by making bird lives family lives.

A final form of anthropomorphism in *Birds through an Opera Glass* involves race and nation. As with the other forms of anthropomorphism, racial traits are used to help readers classify birds. References to race and nation also link taxonomic classification hierarchies to social hierarchies. Hierarchies of race and nation in particular are naturalized via anthropomorphic descriptions of birds. Describing differences among birds as akin to differences among humans enables Merriam to equate species difference with human racial and national difference, a miscomparison that disproportionately magnifies the significance of race and nation. Said another way, the birds of the world are made up of different species, but humans are not. Merriam writes, "In the spring the yellow-bellied woodpecker is a mercurial Frenchman compared with the sober, self-contained Englishmen, his cousins, the hairy and downy" (160). This equation between species and nationality (which seems meant to be humorous) could seem as harmless as the previous comparisons between birds and "little gray-robed monks and nuns," but there is a key difference. Because of such misalignments, Merriam is able to draw from concepts of social Darwinism:

> Birds' bills are their tools,—the oriole's is long and pointed for weaving, the chickadee's short and strong to serve as a pickaxe; but when the nest does not call for a tool of its own the bill conforms to the food habits of the bird, —as the white man's needs are met by knife and fork, and the Chinaman's by chopsticks. (82)

Here, separate bird species are equated, through the simile of beaks to eating utensils ("tools"), with separate human races. The viability of human races, then, is presented as an issue of natural selection. To be clear, Merriam's discussion of beak differences has the simple intention to help birdwatchers tell orioles from chickadees. From our current perspective, though, equating species differences with racial ones confuses the understanding of species while essentializing cultural difference. As casually as Merriam compares birds to "companies of little gray-robed monks and nuns," she aligns different bird species with "white men" and "Chinamen." In doing so, she is expressing common perceptions of human difference at the time. This engagement with the endemic racialized discourse recrudescent in her era of Jim Crow law shows how taxonomic science writing of the kind found in early field guides was inextricably connected with cultural investments in magnifying racial difference.

Merriam's is a field guide in which Redstarts show warbler "blood"

(183) and "everything about [the Robin] bespeaks the self-respecting American citizen" (4). In that *Birds through an Opera Glass* at times comments on and critiques the social order, the global emphasis on species difference (over similarity or relatedness) reifies and is congruent with the cultural impulse to emphasize gender and racial difference (as opposed to similarity) above all else. This impulse is part of what Mary Louise Pratt describes as the Eurocentric "descriptive apparatus of natural history" (1992) that can be said to be reemergent in the genre of the field guide. Because of these types of comparisons, birdwatching is not merely about seeing and sorting animals in the out-of-doors; it does cultural work. The new genre of the field guide and pastime of birdwatching engage in cultural reproduction while reimagining birds as worth protecting.

THE DISCERNING BIRDWATCHER

To invest bird identification with value, Florence Merriam and Mabel Osgood Wright established dramatic tensions between species that are either adored or condemned in their narrative descriptions, heightening the appeal of the bird world as a kind of spectacle of good and evil. Although Merriam proposes that birdwatchers classify all species of birds they encounter, she unabashedly values certain species over others. In *Birds through an Opera Glass*, a birdwatching outing resulting in the identification of preferred species would be better than one yielding less favored species, and Merriam is frank about stating such tastes and priorities.[10] In the 1890s in the United States, in part because of the popularity of field guides, having a taste for birds and bird feathers on one's hat was recharacterized as in "bad taste," while having a taste for watching and identifying birds in the wild became part of middle-class refinement and good taste. It was largely through the use of narrative in early field guides that field-guide authors during this period managed to help sponsor such significant social transformations and reversals in terms of perceived markers of refinement.

According to Wright, species that warrant identification and protection are contrasted with other species that should be "managed" via hunting and other methods of extermination. Either good or evil, everything gets identified by the birdwatcher. Such was certainly the case at

Wright's Birdcraft Sanctuary, where favored birds were made safe through the eradication of predatory birds (and cats) that were trapped and shot in large numbers (Gibbons and Strom 1988, 182). In contrast to guides of the mid- to late twentieth century that cast all or nearly all wild birds as valuable and worth protecting, Wright sees no contradiction in a conservation ethic that prefers certain species over others.

What earns a bird species high standing in Wright's avian hierarchy is often physical beauty (bright colors and elaborate plumage), but value is also attributed to certain behaviors, utility to people, and beautiful songs. Describing the White-throated Sparrow, for instance, Wright unabashedly admits that she prefers it over other sparrows because of its looks. She writes, "This is unquestionably the most beautiful of all the Sparrows, not excepting the great Fox Sparrow, and its rich velvety markings and sweet voice have made it one of the welcome migrants" (Wright 1895, 151). For Wright, a new birdwatcher needs not only learn to tell one species from another, but discern which ones are more attractive (or "welcome"). Furthermore, as this description of the White-throated Sparrow shows, in Wright's era the value of birdsong was on par with or even more important than visual features, with the pleasant sounds of songbirds thought to enliven both the meadow and home.[11]

In another description, domestic virtue is coupled with service to "man" to earn a species a preferred position in Wright's system of discernment: "The Mockingbird is very valiant in the care of its young, and particularly winning and sociable in its relations with man" (77). Martins are another favorite because of their usefulness to people, as the Martin (our Purple Martin) "consumes a vast quantity of evil insects, and these, too, of a larger size and different class from those captured by other Swallows" (126). Which birds are and are not important is something Wright works to educate her readers about, and that value is not intrinsic but derived in reference to people. For instance, stamping out evil (insects) in the world wins an endorsement, whereas being perceived as injurious to livestock earns a species a death warrant. A "beneficial Owl" is one that feeds "chiefly upon mice and other small mammals, beetles, etc., only occasionally eating small birds" (208), whereas raptors, seen as threats to livestock and poultry, were described negatively.

For Wright, a beautiful song is almost always a selling point. In *Birdcraft*, there are birds that sing beautifully and there are "birds with but poorly developed singing apparatus" (47), and the latter do not move up in Wright's hierarchy of taste. Such birds are not simply different from more "avid" singers, as later field-guide authors would describe them, but are less valuable to listeners in need of pleasing birdsongs. In this way, being discerning when it comes to knowing which birds are good singers and which are not is part of leading a life of good taste in tune with the aesthetic virtues of nature. Indeed, *Birdcraft*'s very ordering of species emphasizes this preference for songbirds, an order Wright justifies in her introduction: "[The guide presents the] Song-birds first, and it is to these that you will be first attracted, and, finding many of them familiar, you will be led by easy stages to the Birds of Prey and the Water-birds" (37). By putting the songbirds first, Wright creates for her readers a kind of aesthetically virtuous gateway into birdwatching.

Wright works to improve the reputations of some beleaguered species and families of birds,[12] whereas several species are boldly and violently condemned. How Wright describes the Blue Jay is a graphic example. In contradiction to Merriam, for whom Blue Jays give "color and life," are "brave," and are capable of "cries of almost human suffering," Wright describes the Blue Jay as loathsome, calling the species

> a bird against whom the hand of every lover of Song-birds should be turned in spite of its beautiful plumage and many interesting ways; for the Jay is a cannibal not a whit less destructive than the Crow. . . . Day by day they sally out of their nesting-places to market for themselves and for their young, and nothing will do for them but fresh eggs and tender squabs from the nests of the Song-birds. (177–78)

For Wright, a native species of bird that kills other valued birds is not a necessary part of biodiversity, as contemporary environmental rhetoric might have it, but in need of management. "Cannibalism," oddly figured by Wright as one species eating another (what we would call intraspecies predation), is considered the ultimate offense, blacklisting the visually lovely Blue Jay.

As the passage suggests, Wright also sustains the common dislike of crows that Merriam resists in *Birds through an Opera Glass*. In reference to crows, Wright's language is increasingly violent: "[The Crow] is another bird you may hunt from your woods, shoot (if you can) in the fields, and destroy with poisoned grain. Here he has not a single good mark against his name" (179). As in the many arguments for eugenics, salient in the era, species that are not promoted to proper citizenship status become candidates for extermination. Being a "cannibal," "coward," and "thief" means a bird species lacks the good character and human virtue that would make it deserving of human protection. Although not a field guide for hunters per se, *Birdcraft* reveals how similar the two kinds of guides were during this period. Birdwatchers, in Wright's scheme, need not only learn to identify; they must learn to discern so as to act for and against birds. What may appear to be aesthetic virtues such as beautiful plumage can, in some cases, work to hide the "true" nature of rapacious birds such as the Blue Jay or Crow.

Two other species, both European birds introduced into North America in the nineteenth century, become targets of Wright's scorn in part because of what she characterizes as their immigration status. Gary Alan Fine and Lazaros Christoforides explain that Wright was not alone in thinking of nonnative birds as akin to human immigrants, as English Sparrows were broadly cast as a problem that paralleled "the new immigration" of the late nineteenth century (Fine and Christoforides 1991, 375). In a prefatory essay to the ninth edition of Wright's guide, published in 1917, the English Starling (our European Starling) is described as "now . . . a serious menace both to the summer resident birds, whose nesting sites they appropriate, and to the winter food supply" (xxi). This echoes the way the English Sparrow (our House Sparrow) is referred to as "an absolute and unmitigated nuisance" in the first edition of *Birdcraft* from 1895 (136).

In the case of the English Sparrow, distaste for the species even makes its way into Wright's usually objective, scientific taxonomic description. Classifying the English Sparrow, Wright's bias creeps in when she writes: "*Song:* A harsh chirp" and "*Season:* A persistent resident" (136). Here, Wright's science writing is inflected with a rhetoric of xenophobic condemnation. In *Birdcraft,* Wright claims that "the destruction of the Sparrows, eggs and nests, is now almost universally approved in the United States" (137). By pointing to this section in Wright's guide, I do not mean to suggest that the introduction of European Sparrows did not pose an environmental problem, but instead to show that becoming a birdwatcher for Wright involved more than merely falling in love with birds. The genre of the field guide sponsored protectionism by integrating it within a sometimes hostile philosophy of avian management.

Many more birds are condemned by Wright: "The Cowbird is the pariah of bird-dom, the exception that proves the rule of marital fidelity and good housekeeping" (167); "the Sapsucker is a superbly marked Woodpecker, but its beauty is neutralized by its pernicious habit of boring holes in the tree bark through which it siphons the sap or eats the soft, inner bark" (199); the Great Horned Owl "is the most destructive of Owls and of all birds of prey. . . . This savage Owl also destroys vast quantities of game-birds" (213); the Sharp-shinned Hawk "destroyed between twenty and thirty young chickens" (216) and thus should be hunted; the Cooper's Hawk is "a mischievous harrier of all birds from barn yard fowls to Song-birds, doing by craft what it cannot accomplish by daring alone" (217); and the Bald Eagle, according to Wright, was "unfortunately chosen as the emblem of our Republic, for its noble qualities are in reality either wholly superficial or else imaginary" (221).

As I explain in the following section, Wright has no qualms about expressing such marked condemnations about certain species given the way that strong emotional reactions to birds were essential to her formulation of birdwatching. The rhetoric of taste and discernment serves in her writing to intensify emotional reactions to birds, dramatizing the bird world and encouraging birdwatchers to judge and respond to a bird citizenry that is modeled after the stratified human order. Certain birds are to be loved and protected, others despised as destructive, alien, wicked, or morally depraved.

This rhetoric of taste and preference has long been a part of discerning and aestheticized natural history writing and was equally instrumental in colonial nature illustration concerned with documenting lands, plants, animals, and peoples (Pratt 1992). The authors of late-nineteenth-century guides were open about describing which birds they liked and which ones they did not. This rhetoric of preference is part of a belief system that positions people not only as connoisseurs of good and bad nature, but as environmental managers whose responsibility it is to kill off pestilent species while cultivating and protecting desirable ones. It is a eugenicist way of thinking, common during the period, in which good citizen-birds should be supported while bad citizens must sometimes perish. One way the new genre of the field guide mobilized readers to end the plume trade was by tapping into existing nonegalitarian hierarchies of difference and taste.

Sometimes Merriam's preferences are mild and concerned largely with aesthetics. For example, she writes in *Birds through an Opera Glass:* "The barn swallow is the handsomest and best known of the swallows. . . . What a contrast to the ugly so-called 'chimney swallow'" (Wright 1895, 55). Similar examples abound in her field guide. In emphasizing a preference for one species over another, Merriam develops the requisite for a system of taste and discernment: imbalance. She also taps into the existing discourse of disutility that rendered all birds worthless (and thus commodifiable for hats) in the first place. Although Merriam's reproduction of the rhetoric of disutility is emblematic of a set of assumptions that led to the plume trade in the first place, it is an unresolved contradiction in this and other taste-based writing about birds during the period.

Merriam does work to ameliorate the besmirched reputations of, for instance, crows—"The despised crow is one of our most interesting birds" (10), she writes—but she does not vouch for other reviled species. The flycatcher is bad because it is seen as having no benefit to humans—"All the disagreeable qualities of the flycatchers seem to centre in this bird. His note is a harsh, scolding twitter" (84)—whereas Merriam sees hummingbirds as perfectly lovely and thus steeped in aesthetic virtue. She writes: "[The Ruby-throated Humming-bird] seems like the winged spirit of color" (36). Such preferences for certain species over others (and it is usually the colorful, the small, and the hard to find that Merriam values most) even go so far as to prompt her to joke about restructuring the scientific classes. "The cedar-birds go into pigeon-hole No. 7 [a reference to Merriam's pigeon-hole classification chart, found at the back of the book], the place for 'the waxwings,' etc., and when you have examined them you will feel that they deserve a cubby-hole of their own" (112). Although that is another of Merriam's clever jokes, the preference for beauty in this description and Merriam's field guide on the whole is unreserved. The Snow Bunting and Brown Creeper receive only one or two short paragraphs of descriptive texts, but Merriam's prose overflows for several pages when describing the birds she likes.

Merriam's and Wright's field guides, then, are books that encourage a birdwatching disposition that favors certain birds over others and, indeed, entire families of birds over other families. Such beliefs are set forth as not merely tolerated, but critical to becoming a discerning birdwatcher.

Certain birds, not all of them, are to be sympathized with, admired, sought after in the out-of-doors, and protected, a characteristic of early guides that indicates ways authors like Merriam and Wright leveraged salient biases and assumptions to help sponsor ways of looking at birds through already existent cultural lenses and ideologies. This hierarchization of the bird world by systems of taste and preference is one of the primary inheritances passed down to contemporary systems and formulations of birdwatching.

EMOTIONAL BIRDWATCHING

Wright makes it clear in her field guide that emotion and science are compatible in bird study: "Nature is to be studied with the eyes of the heart, as well as of the microscope, and ever so scanty a knowledge of our feathered brothers helps us to feel that the realms of Nature are very near to the human heart and its sympathies" (1895, 39). Merriam's and Wright's guides, as well as those by their contemporaries Olive Thorne Miller (*Bird-ways*, 1885; *The First Book of Birds*, 1899) and Neltje Blanchan (*Bird Neighbors*, 1897), describe affective, sympathetic relations between birdwatchers and birds.

One way this emotional complex (meant to foster environmental conservation) is constructed is through narrative accounts of favored birds being unjustly and cruelly killed. Another even simpler way involves naming birds to render them emotionally charged objects or catalysts. As Linda Forbes and John Jermier have put it, "Like other conservationists, Wright was distressed by sentiments and practices at the turn of the century that were leading to the widespread devastation of many species of birds" (2002, 459). Overt references to emotion and sentiment (part of the nineteenth-century literary tradition of sentimentalism) exemplify a way early American field guides combine science writing with other popular literary forms. As described in chapter 2, guides published during and after the 1930s moved away from this melding of sentiment and science.

In Wright's introductory chapters, she works to describe how affective bonds of friendship and neighborliness can be fostered between readers and birds.[13] If bird study is done right, Wright claims, "You will discover that you have neighbours enough [in birds], friends for all your moods, silent, melodious, or voluble; friends who will gossip with you, and yet bear no idle tales" (Wright 1895, xv). Moods, in this way of thinking, can be matched up with and elicited by certain species, and some species are described as indispensable when it comes to fostering a particular mood.

Similarly, the wrong birds risk ferrying the viewer/listener off to bad moods. Wright suggests, "Now is the time to study all the best attributes of bird life, the period when we may judge the birds by our own standard, finding that their code of manners and morality nearly meets our own" (12). With such ethical equivalencies in place, affective bonds can be formed between birdwatcher and the appropriate recipient of the birdwatcher's attention. The beneficiary of such an emotional transaction is thought to be both birds (by nature of being protected) and birdwatcher (by nature of drawing value from the birds in the form of good feeling). This kind of birdwatching, using "the eyes of the heart," is emotional, caring, and ultimately comprised of strong sympathetic relations between birdwatcher (as receptive emotional subject) and well-chosen, carefully identified bird (as ethically worthy catalyst and object for feeling).

For Wright, the song of the Wood Thrush, for instance, is capable of "compassing in these few syllables the whole range of pure emotion" (58), emotion that is created through listening to and knowing the thrush's fine song. During the bleakest times of the year, birds are described as integral to emotional salvation: "The Winter Wren is one of the group of tiny birds that enliven December, January, and February" (85). During the summer months, of course, birds are still needed, although no longer to drive away winter blues. Not only are birds worthy and evocative of human emotion, but they are capable of it themselves as "lovable, cheerful little spirits, darting about the trees, exclaiming at each morsel that they glean" (95). On the dark side, a harbinger of doom like the Great Horned Owl can suggest "every form of dark emotion by its voice" (213). In these affective dimensions, the narrative descriptions in *Birdcraft* serve as species descriptions in that the complex narratives aid would-be birdwatchers. At the same time, the descriptions become maps of how to value and feel about birds, incorporating birds into the domestic, moral, and emotional lives of the birdwatcher.

For Wright, birdsongs in particular can be fonts of emotional rejuvenation. She writes, "And best of all, the Hermit Thrush, whose heavenly notes of invocation, if once heard, are never forgotten" (6). The Hermit Thrush is preferred over other thrushes because the species engenders such an ability to "invoke" and, keeping classification in mind, be remembered for that song. Wright's Ruby-crowned Kinglet is another emotional messenger: "In late autumn, even after a light November snow, these cheery, sociable, little birds come prying and peering about the orchard" (70). The Catbird is a "merry, rippling jester, this whirlwind of delightful mockery . . . a companion to the Thrushes with their spiritual melodies" (80). These are some of the birds Wright is quick to point to as her favorites, and in her descriptions of them she takes for granted that knowing birds by their songs is integral to allowing those songs to affect one's emotional state. "Messengers" of nature, filled with delight and energy, have the power to imbue their watchers and listeners with emotional satisfaction while providing one more reason to take up birdwatching and conserve songbirds.

Because tropes of literary sentimentalism were widely used to promote progressive social change in the eighteenth and nineteenth centuries, it is no surprise that field-guide authors turned to sentimentalism to make arguments for bird protection. Stephen Hartnett explains that sentimentalism, despite being commonly seen as simple literary entertainment, had a distinctly political edge. As he writes, sentimental authors have "a fascination with some of the paradoxical dilemmas of modern democracy" (Hartnett 2002, 15). Because of this tradition of using sentimentalism to confront pressing problems, early field-guide authors fused scientific descriptions of birds with sentimental ones to construct birdwatching as a pastime concerned with progressive environmental, if not social, change. Such a mixture of the scientific and the literary was already quite common in natural history writing of the period and helped make bird study seem scientific while aligning bird protection with other progressive social agendas.

Whereas joining birdwatching to ornithology gave birdwatching credibility, joining it to sentimentalism brought with it the problem of being seen by some as less than serious. Understanding the contentious position of sentimentalism ultimately does a good deal to explain why later field guides (published in the twentieth century) moved away from both tropes of sentimentalism. Philip Fisher explains this kind of move in showing how the twentieth century became defined by a "struggle against sentimentality" as reliance on affect, a key element in sentimentalism, came to be seen as dangerous (Fisher 1985, 95).

For Hartnett, sentimentalism in the nineteenth century was both needed and effective because the typical forms of political discourse had become foreclosed and ineffectual. The use of sentimental rhetoric, and thus emotional appeals, "need not be read as compromising its potential persuasiveness. Instead, one may argue that such layered emotional appeals stood a better chance of leading toward political commitment than more traditionally rational (or narrative) appeals precisely because they were so personal, so conflicted, so inescapably woven into the intricate yet always fragmentary experiential fabric of everyday life" (Hartnett 2002, 5). Sentimental narratives had contributed to battles for abolition, women's suffrage, more socially acceptable roles for women, and, as early American field guides to birds demonstrate, new and imaginative strains of environmentalism. Hartnett writes that the "prose that we find emotionally overdetermined left our ancestors productively shocked and emotionally tangled; so much so that they then began to consider dramatic political and economic changes that, prior to the emotional cathexis prompted by sentimental texts, seemed absurd or imprudent" (Hartnett 2002, 5). Florence Merriam and Mabel Osgood Wright were two early field-guide authors who transformed sentimental descriptions of trapped, abused, and neglected female heroines, children, and slaves—common tropes of nineteenth-century sentimental fiction (Fisher 1985; Tompkins 1985)—into "overdetermined" descriptions of trapped, abused, neglected, and hunted birds. It was all meant to make birds seem worth saving.

As Mark Barrow writes, "By reinforcing an emotional and aesthetic bond between humans and birds, birdwatching helped convert ordinary citizens to ardent conservationists" (1998, 156). At the same time that the scientific discipline of ornithology was taking shape and as the mass extinctions of countless bird species seemed inevitable, rhetorical tactics that would have been culturally familiar, such as working to engender sympathy for fragile and persecuted creatures, were mixed with science and put to use in the first American field guides.

This formula seemed to appeal to readers, but only as long as sentimentalism was in favor. By tapping into the existing literary tradition of sentimentalism, such appeals came into conflict with the tenets of perceived objectivism central to early American ornithology. In a sense, sentimental field guides quickly became incompatible with birdwatching as an objective amateur science. By relation, overt appeals to bird preservation, which had been sentimental, were incompatible as well. Sentimental field guides dramatized what were often referred to as "bird lives," putting birds on equal terms with other persecuted human subjects. In such dramatizations, marked preferences for certain species over others be-come apparent, but this sort of thing was dissonant with ornithological practices of giving all species equal weight in scientific descriptions. In that the environmental politics of birdwatching field guides emerged as highly reliant on sentimentalism, the problem became that the foundation of bird conservation in early field guides to the birds of North America was untenable by the 1920s and '30s. As sentimentalism fell away in field-guide literature, so too did overt appeals to conservationism.

CONSTRUCTING BIRDWATCHING AS CONSERVATIONIST

This chapter focused on two of the earliest field guides to the birds of North America, Merriam's *Birds through an Opera Glass* (1889) and Wright's *Birdcraft* (1895), to explain how the genre repurposed existing literary, cultural, and scientific traditions to intervene in the overkilling of birds by sponsoring a new hobby. To make bird appreciation and identification appealing, these field guides grafted amateur-science practices of taxonomic identification onto already existing systems of beliefs about anthropomorphism, citizenship, ethics, taste, and emotion. This connection, between taxonomic discourse and late-nineteenth-century dominant social orders, worked in many ways because both systems were hierarchical. The prominence of multiple forms of anthropomorphism and a rhetoric of discernment in these early guides reveals the comfort their authors had inflecting birdwatching with social commentary and other tropes from natural history, children's literature, and sentimental fiction. Through its multiple forms of appeal, the new genre of the field guide proved tremendously successful in helping save birds through the formation of the new textually mediated pastime.

In part, by beginning this discussion of field guides and binocular vision with two of the first guides to North American birds, I have sought to recover and understand the rhetorical politics of science, sentiment, emotion, and social issues that helped spawn the pastime. Even though the type of field guide discussed in this chapter would lose popularity and salience in the coming decades, many of the attitudes and beliefs present in the first guides are still prominent among birdwatchers.

With a range of compelling and at times contradictory ideas, Merriam's and Wright's guides inserted themselves into daily life and helped expand the available range of known daily activities. By the 1890s, it was possible to pick up the quasi-scientific, quasi-domestic practice of birdwatching. The genre of the field guide and new pastime of birdwatching were not alone in working to save birds—the American Ornithologist Union's Committee for the Protection of North American Birds has been cited as critical to avian conservation (Barrow 1998, 106–7)—but the guides were one instrumental force in bringing about social and attitudinal changes at the end of the nineteenth century.

It is important to remember that although *Birds through an Opera Glass* and *Birdcraft* included a number of initiatives and served multiple goals, the main organizing goal under which everything else in the guides was subsumed was species identification. These books assume that anthropomorphism, discerning birdwatching, and strong emotional feelings about birds can only be achieved once a species is identified and named. As with the racialized subjects birds are often compared to, these guides assume that birds first need to be identified, classified, and named before any substantive relations with them can be formed. This axiomatic and indispensable role of identification is what is most taken for granted in these guides and is taken up by subsequent field-guide authors. Birdwatching began and continued as a textually mediated environmental pastime based on the pleasure and satisfaction of naming birds, and that act of naming was intended to help foster environmental conservation early on.

In addition to contributing to the diminishment of the plume trade at a moment of environmental crisis, the new field guides sponsored a long-lasting hobby. This hobby is based on the use of what theorist Yrjö Engeström calls mediating artifacts (2006); in the case of birdwatching, these artifacts are opera or field glasses, bird notebooks, lists, and field guides. Accomplishing identification through activity involving mediating artifacts became a defining feature of birdwatching encounters. Somewhat unlike other emerging genres that have received scholarly attention (T. Miller 1997; Popken 1999), the genre of the field guide was strategically crafted, published, and circulated to form, enable, regulate, and maintain a new everyday pastime of environmental importance. In this sense, the genre of the field guide began as an intervention aimed at promoting conservation. As an account of genre formation, then, it is a story of a genre being pushed onto the scene in an attempt to restructure everyday life in ways that promote environmental protection.

In recent decades, it has often been assumed that *confrontation* is needed to bring about environmental conservation. Members of Greenpeace put their lives on the line to save whales, confronting whaling ships in rubber dinghies; members of Earth First! sit in trees and block roads to protect old-growth forests (DeLuca 1999); others take to the streets in various forms of direct-action and theatrical protest (DeLuca and Peeples 2002; Pezzullo 2003). But in the 1880s and '90s, Wright and Merriam attempted to sponsor environmental conservation with a very different, noncombative approach that encouraged readers to watch, appreciate, and even love a maligned class (*aves*) of species. The fundamental restructuring of everyday life that field guides attempted gently to promote was intended, of course, to cease commerce in birds and bird feathers. The guides did a lot more than that, though, helping establish a new leisure-time activity in everyday life with an usually high connection with field guides. In defining the parameters of the hobby, field-guide authors made their books indispensable.

In the twentieth century, anthropomorphism and preference for certain species over others, as well as overt mention of the value of emotional connections with birds, would become minimized and veiled in guides to North American birds. The genre of the field guide would soon be produced on offset presses, helping make field guides image-rich and decidedly visual, but it was the late-nineteenth-century field-guide authors and their extensive use of narrative that leveraged complex appeals to both sympathy and classificatory identification to help put an end to the widescale destruction of birds, foster a new pastime of birdwatching, and establish field guides as a central, governing, indispensable genre capable of validating identificatory encounters with living birds.

CHAPTER TWO

Nuisance Birds, Field Guides, and Environmental Management

The Bald Eagle is currently known as a patriotic symbol within the United States, but the bird was once despised for its work habits and killed by the thousands as a bounty bird in Alaska. Crows still have a bad reputation. Today they are hunted with few regulations across North America even as research shows diminishing populations of American Crows (due to West Nile virus) and evidence of remarkable crow intelligence. Mute Swans, once imported to North America because of their beauty, are being recast as pests by the U.S. Fish and Wildlife Service and targeted for mass extermination. These examples are just a few of the more than one hundred nuisance birds in North America. Although many species of birds are seen negatively and treated accordingly, since the 1930s such birds have been described in field-guide literature with little or no mention of their divisive positions as nuisance birds.

As chapter 1 describes, Florence Merriam and Mabel Osgood Wright were strong advocates for birdwatching and bird preservation, but it did not mean that they professed the love and protection of all birds. Instead, like other late-nineteenth-century bird advocates, Merriam and Wright were harsh critics of multiple species of birds, particularly those they saw as a nuisance; managing certain species was part of their conservation ethic.[1] The complexity of this outlook has often been ignored in favor of seeing such writers simply as "bird lovers." Linda Forbes and John Jermier try to reconcile Wright's mixed feelings toward birds, however, by characterizing Wright as an environmentalist who "espoused a philosophy of natural rights for all living things" (2002, 459) while practicing what they call a nativist model in managing her bird sanctuary (464).

But Wright's antipathies extended beyond nonnative species; her brand of conservationism was not as much a philosophy of natural rights toward all living things as a natural philosophy based on preference for some species and open antipathy toward others. In her field guide, Wright calls the American Crow, a native species to North America, "a coward, with a hoarse voice and disagreeable manners added to a most offensive, crouching personality hiding a world of cheap craft" (1895, 179). Wright's House Sparrow gives "only his ugly chirp in the place of [the] songs [of our most familiar birds]" (136–37), and the native Yellow-bellied Sapsucker is described as a "pernicious" "evil-doer" (199).

At the end of the nineteenth century, birds that were seen as a nuisance or construed as varmints by farmers, ranchers, trappers, and fishermen included any species that was thought to damage crops or undesirably prey on other animals. Crows, cormorants, gulls, and what were called "birds of prey" were widely hated for these reasons, and these negative feelings fueled the aggressive management of nuisance birds and varmints and vice versa.

By the 1930s, with the publication of Roger Tory Peterson's *A Field Guide to the Birds* (1934), field guides took a decided turn in becoming technical and visual. Authors focused more on the intricacies of field identification and left out any mention of conservation or their feelings about individual species. The move away from overt discussions of conservation can in part be explained by the lessening of the environmental crisis that early field-guide authors had responded to, along with the passage of the Migratory Bird Treaty Act into U.S. law in 1918, but it is not as if the 1930s marked an end to strong sentiments about birds and a need to protect them. Rather, those strong feelings were simply no longer mentioned in birdwatching field guides.

Management philosophy continued to contend, for instance, that there were good birds and bad birds; in 1918, one such "bad bird" was the Bald Eagle. By the 1930s, nuisance birds were described in field guides as just as important for birdwatchers to identify as any other bird. This seemingly egalitarian perspective produced what I contend is an abstracting lens that tries to make birdwatchers complicit with the changing and often problematic landscape of avian environmental protection and management. Binocular vision in part renders invisible ongoing and lethal forms of avian management.

The 1970s and '80s marked the beginning of a significant rise in the numbers of nonnative bird species in North America, resulting in more species of birds than ever before being categorized and treated as nuisance birds, but field guides continue to ignore the negative perception of these birds and how they are managed. Because every known species of wild bird has value as an identifiable object for the birdwatcher to look at and name, nuisance birds have largely been imagined as uncoupled from the cultural and legal frameworks that involve and affect such species.

Seeing birds as somehow separable from culture and policy has become a critical part of what I call binocular vision. It means that technical field guides have, since the 1930s, turned a blind eye on several of the more divisive cultural and scientific debates over nuisance birds to instead focus on a uniform, technically minded process of identification, thereby fostering a sanitized view of how nuisance birds in particular are perceived and treated. In addition, the overt conservationist beginnings of birdwatching have been toned down while birdwatching continues to be characterized and promoted as green by mainstream environmental groups such as the National Audubon Society, American Bird Conservancy, and National Geographic Society. Each of these groups endorses its own particular birdwatching field guides to raise money for their organizations and brand birdwatching as a green pastime. Although birdwatching certainly relies on saving some birds, as I will show, it also continues to rely on killing others.

This chapter explores representations of four nuisance birds—the Bald Eagle, Mute Swan, gulls, and crows—to explain how some of the birds birdwatchers seek can be seen through cultural, legal, and environmental lenses that are much more robust and revealing than those provided by the standard tools birdwatchers use. In this way, the chapter traces the continuation of some of the strong negative sentiments and reactions to "bad birds" raised in chapter 1, asking, *What happened to nineteenth-century antipathies to certain birds?*

I begin with a bit of textual history about field guides in the early decades of the twentieth century, a period when guides moved away from the kinds of sentimental and narrative nature writing described earlier to become thoroughly technical, visual manuals. With this change came the promotion of the abstracting lens that sees birds as identifiable objects in the landscape that can and should be removed from both environmental and cultural contexts so as to be identified by the birdwatcher. After sketching how the textual history of field guides contributed to this way of seeing, I argue that field guides consent to a range of status quo, inconsistent management laws and practices. Of the more than one hundred species of birds seen and treated as a nuisance in North America,[2] the four examples discussed reveal how deeply enmeshed nuisance birds are in changing U.S. environmental laws, nationalism, environmental management philosophies, belief in the importance of biodiversity, hunting laws, and the popular environmental imagination.

Historians of birdwatching have described the many contributions birdwatchers have made to avian conservation in particular and the U.S. environmental movement more generally. This chapter complements those histories, showing how field guides have been crucial sponsors of conservation efforts while existing in a cultural and environmental context in which not all birds are seen or treated equally. The designation of which birds are and which are not labeled a nuisance has been highly unstable, with various species of birds moving in and out of the category. Desirable species as well as vermin continue to be part of the birdwatching landscape.

TECHNICAL FIELD GUIDES AND BINOCULAR VISION

It is certainly not the case that, at one point in time, field guides suddenly became technical and relied primarily on images instead of narrative to represent birds; nor is it the case that there is a single moment when field-guide authors stopped overtly condemning nuisance birds. Instead, a new kind of technical, visual, noncondemnatory field guide emerged gradually in the early decades of the twentieth century, coming firmly into place by the 1930s.[3] As discussed earlier, field guides published at the end of the nineteenth century provided readers with information about birds couched in narrative that was frequently anthropomorphic and particularly attuned to the emotional register of birdwatching.

Several of these first field guides continued to be popular into the twentieth century, with Wright's 1895 *Birdcraft* going through nine editions, the last of which was released by Macmillan in 1936. At the same time that Merriam's *Birds through an Opera Glass* (1889) and Wright's *Birdcraft* (1895) were first being read, there also existed a different kind of bird literature that was oriented toward describing birds from much more taxonomic and anatomical perspectives. Two detailed ornithological manuals had been published in the 1870s—Elliot Coues's *Key to North American Birds* (1872) and Spencer Baird, Thomas Brewer, and Robert Ridgway's *History of North American Birds* (1874)—and ornithologist Frank Chapman published his own somewhat technical field guides in the 1890s.[4] Chapman's approach, which rivaled early, narrative field guides, became much more predominant by the 1930s.

The simplest way to date the emergence of modern field guides, and the way most historians of birdwatching have done it, is to emphasize the importance of Roger Tory Peterson's *A Field Guide to the Birds*, first published in 1934. Much has been written about the significance of Peterson's 1934 guide, with the consensus being that the slender book had a remarkable influence on the development of birdwatching (Barrow 1998, 179; Moss 2004, 133–36; Weidensaul 2007, 190–218). Stephen Moss, author of *A Bird in the Bush: A Social History of Birdwatching*, captures this view when he writes that "in the three quarters of a century since its publication, the 'Peterson Guide' has had more effect on the growth and development of birdwatching as a mass participation pastime than any other book in history" (2004, 133). Most scholars tend to agree that Peterson did not invent the modern, technical field guide as much as bring together existing trends in field-guide design and bird identification at the time, most notably Ludlow Griscom's concept of the field mark (Barrow 1998, 178–79). Field marks are diagnostic features meant to help render a species unique and identifiable for the birdwatcher, and although John Lynch and Michael Law have shown that not all species can be identified using a visual field-mark method (1999), Peterson was an unequivocal advocate of the approach. Field marks, in a sense, make bird identification possible with the help of a visual guide, thus ensuring the perpetuation of the guides themselves.

Taking the case of a mature Bald Eagle, such things as the size or diet of the bird would not function as field marks because these features are shared by several other species, but the Bald Eagle's vocalizations, plumage color (particularly on the head), and behavior combine to form a cluster of distinguishing field marks. In Peterson's 1934 description of the Bald Eagle, field marks are used to render the bird unique and identifiable:

> **Bald Eagle.** *Haliætus leucocephalus.*
>
> This, the typical Eagle of the East, needs little description. The adult, with its great size and *snowy-white* head and tail, resembles no other bird of prey.
>
> The immature bird is ducky all over. Melanistic Buteos (black Rough-legs, etc.) are much smaller, with more or less gray and white under the wings. Some of the sight records of Golden Eagles in the East could doubtless be assigned to young birds of this species, although there are distinct points of difference that should render such mistakes inexcusable.

Peterson's eagle is not necessarily good or bad, regal or slovenly; it is simply a composite of field marks.

This example of the Bald Eagle reveals how Peterson moved away from value-laden condemnation or approval of various species to fault the novice birdwatcher who would not be able to identify such an "easy" bird. I turn to the example of the Bald Eagle because, at the time Peterson was writing his field guide, it was a species that was disliked by many. This sentiment came out of a long history. Benjamin Franklin referred to the Bald Eagle as a bird of "bad moral character" (Franklin 1784), and Wright seconded Franklin's critique in *Birdcraft*, calling the Bald Eagle "unfortunately chosen as the emblem of our Republic," "an inveterate bully," and "cowardly" as a parent (1895, 221). In 1917, fifteen years before the publication of Peterson's field guide, a bounty (discussed in more detail below) was placed on the Bald Eagle in the territory of Alaska. But Peterson's text makes no value judgments about the eagle, does not weigh in on the debate surrounding the bird, and never mentions the bounty.

Although Peterson begins by saying that the Bald Eagle "needs little description," he then carefully explains how certain forms of the species, particularly immature Bald Eagles, might be distinguished from similar-looking raptors. The "Melanistic buteos" Peterson mentions are raptors that, due to plumage variation, can look a lot like an immature Bald Eagle, so he mentions this group as similar but "small, with more or less gray and white under the wings." Because the Petersonian field marks of the "*snowy-white* head and tail" distinguish adult Bald Eagles from other raptors without exception, Peterson does not linger on the adult birds. At the end of the description, when Peterson plugs his field-mark system as consisting of "distinct points of difference . . . [that] should render such mistakes unexcusable," he is arguing for the viability of that system. Field guides work, he believes, because they describe and illustrate only those features of a species a birdwatcher needs to know to distinguish one species from another. No more is needed.

Although individual members of a species can differ quite a lot, Peterson ignores this kind of variation. Also of no concern are cultural debates surrounding a species (like the bounty in Alaska) that might adversely affect a population's longevity. Peterson's guide makes a clear statement that birdwatching is about focusing on conclusive forms of difference at the level of species, and it was this form of birdwatching that was endorsed, encouraged, and in part made popular by the success of *A Field Guide to Birds*.

Describing Peterson's guide as unconcerned with cultural and environmental debates involving Bald Eagles does not mean, however, that technically oriented field guides are without agenda or bias. Most obviously, technical field-guide literature privileges scientific understandings of nature developed within the discipline of ornithology, thus subordinating the importance of more emotional or cultural ways of seeing and thinking about birds.

Technically oriented field guides promote seeing all birds as identifiable through a process that visually separates each bird from its immediate environment and from the larger cultural debates each species is a part of. Similarly, by the 1930s technical field guides developed into books that are decidedly biased toward the visual aspects of birds over the sounds they make, a topic taken up in chapter 4. So, even though the 1930s marks the time when field guides no longer weighed in on debates over whether a certain species was revered or seen as a nuisance, many other forms of bias still existed.

Following the enormous success of Peterson's field guide (published during the Great Depression, the first two thousand copies printed sold out in the first week [Moss 2004, 134]), narrative guides such as Wright's *Birdcraft* went out of print. Peterson's field guide quickly became *the* field guide to birds in the United States, and it maintained that status for more than twenty years. Not until the publication of Chandler Robbins, Bertel

Bruun, and Herbert Zim's popular 1966 guide, *Birds of North America: A Guide to Field Identification,* did Peterson's guide face any real competition in the marketplace.

The newer guide, often referred to by birdwatchers simply as "the Golden guide" (named after its publisher), incorporated several additional technical features such as sonograms of birdsongs, comparative images, and range maps. These innovations were taken up and responded to in turn in revised editions of Peterson's guide as well as in a host of other similarly technical field guides: the National Geographic Society's *Field Guide to the Birds of North America* (Allen and Hottenstein 1983), the *National Audubon Society Field Guide to North American Birds* (Udvardy and Farrand 1994), and David Allen Sibley and the National Audubon Society's *The Sibley Guide to Birds* (2000). All these field guides, of which there are many more, rely on a close relationship between technical descriptions of field marks and diagnostic images of birds. These books are also decidedly visual in orientation, featuring multiple images of each species to enable comparisons and highlight visual features.

As subsequent field guides were published and Peterson's guides were revised, written descriptions became increasingly precise. In his 1934 guide, Peterson said that the Bald Eagle "needs little description," but by his 1961 second edition of *A Field Guide to Western Birds,* Peterson includes the following details:

BALD EAGLE *Haliætus leucocephalus* 30–43 p. 75

Field Marks: Spread 6½– 8 ft. The national bird of the U.S., with its *white head* and *white tail,* is "all field mark." Bill of adult yellow. Immature has dusky head and tail, dark bill. It shows whitish in the wing-linings and often on the breast.

Similar species: (l) Golden Eagle is frequently confused with immature Bald Eagle (which see). (2) Black Buteos (Rough-leg, etc.) are much smaller, with smaller bills.

Voice: A harsh, creaking cackle, *kleek-kik-ik-ik-ik,* or a lower *kak-kak-kak.*

Where found: Alaska, Canada, to s. U.S. **West:** *Breeds* from Aleutians, nw. Alaska, Mackenzie south locally to n. Baja California, c. Arizona, w. New Mexico; most numerous in coastal Alaska, B.C. Some withdrawal in winter from colder n. areas.

Habitat: Coast, lakes, rivers. Nest: A bulky platform of sticks in tall tree, cliff. Eggs (2–3) white.

• **EAGLES Subfamily Buteoninae** (in part). Eagles are distinguished from buteos, to which they are related, by their much greater size and proportionately longer wings. The powerful bill is nearly as long as the head. **Food:** Golden Eagle eats chiefly rabbits and large rodents; Bald Eagle, chiefly dead or dying fish.

BALD EAGLE **M 173**
Haliaeetus leucocephalus 30–43" (75–108 cm)
The national bird of the U.S. The adult, with its *white head* and *white tail*, is "all field mark." Bill yellow, massive. The dark immature has a dusky head and tail, dark bill. It shows considerable whitish in the wing linings and often on the breast (see overhead pattern, p. 167). Spread 7–8 ft.
Similar species: See Golden Eagle.
Voice: A harsh creaking cackle, *kleek-kik-ik-ik-ik*, or a lower *kak-kak-kak*.
Range: Alaska, Canada, to s. U.S. **East:** Map 173. **Habitat:** Coasts, rivers, large lakes; in migration, also mountains.

GOLDEN EAGLE *Aquila chrysaetos* 30–40" (75–100 cm) **M 174**
Majestic, the Golden Eagle glides and soars flat-winged with occasional wingbeats. Its greater size, longer wings (spread about 7 ft.) set it apart from the large buteos. *Adult:* Uniformly dark below, or with a slight lightening at the base of the obscurely banded tail. On the hindneck, a *wash of gold*. *Immature:* In flight, more readily identified than the adult, showing a *white flash in the wings* at the base of the primaries, and a white tail with a broad dark terminal band.
Similar species: (1) Immature Bald Eagle usually has some white in the wing linings and often on the body. Tail may be mottled with white at base; but is not distinctly banded. (2) Black morph of Rough-legged Hawk is smaller, has more white under wing.
Voice: Seldom heard; a yelping bark, *kya*; whistled notes.
Range: Mainly mt. regions of N. Hemisphere. **East:** Map 174. **Habitat:** Open mts., foothills, plains, open country.

■ **OSPREYS Family Pandionidae.** A monotypic family comprising a single large bird of prey that plunges feet-first for fish. Sexes alike. **Range:** Occurs on all continents except Antarctica. **No. of species:** World, 1; East, 1.

OSPREY *Pandion haliaetus* 21–24½" (53–61 cm) **M 175**
Our only raptor that plunges into the water. Large (spread 4½–6 ft.); blackish above, white below. Head largely white, suggesting Bald Eagle, but with a *broad black cheek patch*. Often flies with an angle in the wing, showing a black "wrist" patch below. Hovers on beating wings and plunges feet-first for fish.
Voice: A series of sharp, annoyed whistles, *cheep, cheep*, or *yewk, yewk*, etc. Near nest, a frenzied *cheereek!*
Range: Almost cosmopolitan. **East:** Map 175. **Habitat:** Rivers, lakes, coasts.

158

Figure 2.1. Field marks are emphasized in the images using arrows. This two-page spread shows entries for the Bald Eagle, Golden Eagle, and Osprey. Illustrations from *A Field Guide to the Birds East of the Rockies*, 4th ed., by Roger Tory Peterson. Copyright © 1980 by Roger Tory Peterson. Reprinted by permission of Houghton Mifflin Harcourt Publishing Company. All rights reserved.

This description recognizes a much wider distribution of the species and more specifically addresses such field marks as vocalization, habitat, nest type, and eggs. The design of field guides also changed over this period, with text and images being either combined on one page or presented on facing pages (figure 2.1). With multiple indexes and color tabs, field guides became easy for the birdwatcher to flip through quickly in search of an unknown bird.

In figure 2.1, facing pages from Peterson's 1980 guide describe the Bald Eagle, Golden Eagle, and Osprey, and we see the role of what I have called abstraction in this kind of representation. Following the conventions of natural history illustration (Blum 1993, 6), exemplars of each species are either featured before blank backgrounds or near generic-looking trees and landscapes; birds are not a part of the worlds around them as much as removed from them for purposes of easy identification. Arrows point to field marks in the images, suggesting precise attention to seeing birds as composites of field marks. It is not a lens that looks for commonality or connectedness between species, nor is it a way of seeing birds as enmeshed in environmental debates. Instead, it is an abstracting lens obsessed with noting difference in the service of taxonomic naming. The images of birds in field guides are simple and almost always in profile; such poses maximize the display of visible field marks.

Describing the contribution that Peterson's 1934 guide made to birdwatching, historians Felton Gibbons and Deborah Strom describe Peterson as having "invented an obsessive game. He supplied the player with neat bundles of clarified information about the differences among species, and the participant after sufficient observation identified a bird and so added it to his life list. The game, packaged by Peterson to be played by all comers, has become more and more popular ever since" (1988, 300). In the sections that follow, I show how this characterization of birdwatching as a harmless, fun game ignores the extent to which, especially in the case of nuisance birds, the objects of study that birdwatchers seek are hotly contested and in some cases lethally managed.

For the game of birdwatching to be played, avian biodiversity must remain and extinction must be avoided. What is of less concern within the frame of this game is exact numbers of each species, so several thousand superabundant gulls can be culled in one area with no significant damage done to the viability of a gull species. This game, as a form of environmental practice, involves a commitment to management that looking for "neat bundles of clarified information" fails to recognize.

BALD EAGLE: FROM BOUNTY BIRD TO NATIONAL SYMBOL

In birdwatching field guides, the Bald Eagle is known as a large bird with a white head and white tail. Since the emergence of field guides in the 1880s, these features have not changed. What has changed, however, is how the Bald Eagle is perceived and treated. The Bald Eagle is currently well known as the national bird of the United States and as a species with a special, protected status, but that was not always the case. The Bald Eagle was disliked by many in the nineteenth century and hunted as a bounty bird in Alaska for part of the twentieth. Only in the 1960s and '70s were Bald Eagles significantly protected by U.S. environmental laws in ways that contributed to their recovery.

To supplement the previous discussion of field-mark-oriented descriptions in field guides, this section delves into the case of this nuisance-bird-turned-national-icon more closely, discussing the roles of bounty hunting, science, law, and environmental nationalism in the near demise and ultimate recovery of the species. Since the Bald Eagle was selected as the U.S. national bird, it has been a flash point for varying sentiments, and those feelings have been critical to how the bird has been seen, treated, and managed.

Dating back several centuries, the Bald Eagle has been regarded as a sacred bird in several American Indian cultures (Curtin 1923, 184; Laubin and Laubin 1989, 231). After being selected as the national bird of the United States in the 1780s, the eagle became a popular symbol in American art and emerged as "the dominant image [of the country] from the Revolutionary to the Civil War" (Lubbers 2000, 18). How, then, starting in 1917, were Bald Eagles hunted as bounty birds in Alaska (Alaska Department of Fish and Game 1994)? The first thing that allowed the Alaska bounty to take place was a steady stream of criticism of the Bald Eagle, starting in the late eighteenth century.

As mentioned, Benjamin Franklin disagreed with the decision to make the Bald Eagle the national bird, calling it a bird of "bad moral character" and preferring the more respectable Wild Turkey over the eagle (Franklin 1784). Franklin's rejection of the Bald Eagle was on ethical grounds: he frowned on having a bird that did not always "earn" its food and thus positively represent the nation. Franklin's critique persisted a hundred years later when, in *Birdcraft,* Mabel Osgood Wright referred to the eagle as "unfortunately chosen as the emblem of our Republic" and "cowardly" in its role as a parent (1895, 221). The character of the Bald Eagle had long been assailed, and this negative view combined with strong forms of economic protectionism to allow lawmakers to institute the Alaskan bounty.

In the early decades of the twentieth century, fox farming and salmon fishing were growth industries in Alaska, and Bald Eagles were thought to prey on both fox and salmon (Alaska Department of Fish and Game 1994). Instead of the older perception of the Bald Eagle as a bird that did not earn its living, now the idea was that the eagle did so too well and in ways that damaged burgeoning industries. As a result, in 1917 a bounty for Franklin's bird of "bad moral character" was instituted in the territory of Alaska.

The bounty was initially set at fifty cents for each Bald Eagle killed, and by 1953, when the bounty was ultimately repealed, more than 100,000 birds had been slain for a price (Alaska Department of Fish and Game 1994). In a testimonial in Bruce E. Beans's *Eagle's Plume: The Struggle to Preserve the Life and Haunts of America's Bald Eagle* (1997), one gets a sense of how prominent bounty hunting for Bald Eagles was in Alaska in the years between 1917 and 1953:

> "But shooting eagles was a big business," [John] Schnabel recalls. "I had friends that made five, six hundred dollars a year. For two dollars each they were killing two, three hundred birds apiece." Ross Hevel, who operated a barbershop[,] . . . also served as a federal commissioner for the territory of Alaska. To prove they had earned a bounty, hunters presented Hevel with two eagle talons; Hevel authorized a voucher and dumped the talons in a 55-gallon drum that sat outside of his shop. "By springtime it was full of eagle feet," Schnabel remembers. "He'd haul it up to the garbage dump and throw them away." (257)

The image of barrels filled with eagle claws captures how an animal species can be transformed, through the mechanisms of a bounty, into a disposable commodity. In his 1934 field guide Peterson called the Bald Eagle "the typical Eagle of the East, needs little description," but, in fact, a good deal was not typical about the eagle at that time. In Alaska, Bald Eagles were being reduced to the exchange value of their talons.

The bounty on Bald Eagles in Alaska was by no means unusual; animals branded a nuisance have been dealt with in this way in North America since the eighteenth century. Bounties work by enlisting the citizenry to accomplish the environmental management policies of the state, and, as such, bounties purchase the support of the citizenry to bolster an aggressive approach to management. In other words, bounty hunting does more than pay hunters to kill animals; bounties construct a wide set of relationships and ways of thinking about nature. Primarily, bounties make killing bounty animals tantamount to performing a kind of justice. Within the structure of bounty hunting, the bounty hunters are also absolved from deciding if a species is a nuisance or not, as the authority establishing the bounty has already made this determination.

In the nineteenth century in the United States, bounties for birds covered a broad class of species known as "birds of prey,"[5] with such bounties encouraging hunters to kill hawks, eagles, and predatory owls in exchange for payment. Most of these bird bounties were motivated by the belief that bounty birds posed some economic threat to farmers or ranchers, and, in this way, bird bounties can be seen as early forms of agricultural subsidies. Old antipathies toward birds helped reinforce the bounties and vice versa, creating a belief that killing bounty birds meant performing legal and ethical forms of justice. Furthermore, the relative pittance offered for single bounty birds (two or three cents per bird in some cases) represents individual members of a species as close to valueless and, at the same time, worth more dead than alive. Bounties can also be seen as hunter subsidies, allowing hunters to "earn their keep" through hunting in an era when it was increasingly hard to do so. Birds of prey were not alone; in addition, crows, House Sparrows, Blue Jays, and European Starlings have all had bounties placed on them in the United States (Born Free 2008).

To put bird bounties in context, it is worth noting that such bounties were part of a larger bounty system for rodents, snakes, and large predators. In the twentieth century in the United States, as Donald Worster describes, the Bureau of the Biological Survey (founded in 1905) largely took over the job of killing large predators such as wolves and bears. This agency, he notes, was predicated on a Progressive belief that "nature as well as society . . . harbors ruthless exploiters and criminals who must be banished from the land" (Worster 1994, 265).

Although the bounty on Bald Eagles was a legal way to encourage the killing of a certain species, just one year after that bounty law was passed in Alaska, the Migratory Bird Treaty Act of 1918 between the United States and Canada was signed into U.S. law (Dorsey 1998, 230). This act signaled how law could alternatively be used to destroy or to protect wild birds. The language of this treaty, originally signed between the United States and Great Britain (acting for Canada), shows it to be aimed primarily at ending trade in wild birds and feathers. As a U.S. law, however, it had no legal bearing on the bounty on eagles in what was then the territory of Alaska (Congressional Budget Office 2004).

In 1940, the passage of a more specific law, the Bald Eagle Protection Act (Bean and Rowland 1997, 93), protected the Bald Eagle in all U.S. states, but this law still did not extend as far as the territory of Alaska.

Passage of the Bald Eagle Protection Act signals a rising concern, though, about the viability of the Bald Eagle as well as revived linkages between the eagle and the nation. It also signals a first step in transferring those concerns into law.

Mark Barrow, in a detailed account of birds of prey during the U.S. interwar years, writes:

> The [Bald Eagle Protection Act], which countered a tradition of treating the eagle and other birds of prey as pariahs, was actually the culmination of a long campaign. For two decades preceding the act's passage, a network of bird enthusiasts had struggled to repair the tarnished reputation of predatory birds, to fight bounty laws aimed at reducing their numbers, and to secure legislation to protect them from continued harassment at the hands of farmers, ranchers, sportsmen, and others who considered them "vermin," fit only for systematic obliteration from the landscape. Although the Bald Eagle Protection Act fell short of achieving its supporters' ambitious aims, it nonetheless stands as a landmark piece of federal wildlife legislation that has yet to gain the historical attention it deserves. (Barrow 2002, 69)

Ultimately, in anticipation of Alaska being made a state in 1959, the Alaskan bounty on Bald Eagles was repealed. And with subsequent revisions to the Bald Eagle Protection Act in 1962, 1972, and 1978 (Bean and Rowland 1997, 93–97), Bald Eagles have been protected with increasingly severe punishments. Legal protection for the Bald Eagle expanded again with the passage of the Endangered Species Act of 1973.[6]

The Endangered Species Act is of importance to the Bald Eagle because, unlike other forms of legislation protecting birds, the law allows for the protection not only of birds, nests, feathers, and eggs, but for the protection of an endangered species' *habitat.* The general intention of the Migratory Bird Treaty Act of 1918 had been to disrupt trade in wild birds; the general intention of the Bald Eagle Protection Act was to stop the killing of Bald Eagles, mainly by ranchers; but the intention of the Endangered Species Act was to preserve endangered species from extinction, no matter what imperiled them, thus allowing for broader measures of protection.

Before the passage of the Endangered Species Act, U.S. law did yet not speak immediately to what was becoming, in the 1940s and '50s, the greatest threat to Bald Eagles and other raptors: pesticides. Rachel Carson's poetic treatise *Silent Spring*, published in 1962, launched a new wave of criticism of the widespread use of DDT. *Silent Spring* raised awareness about the ill effects of pesticides and insecticides on the eggs of numerous raptors and has been widely credited for raising environmental awareness on multiple levels (P. C. Murphy 2007). In *Silent Spring*, Carson argues for protecting all species of wild birds, making broad claims for avian protection. She writes, "From all over the world come echoes of the peril that faces birds in our modern world" (Carson 1962, 122). Carson also provides data showing that Bald Eagle populations had fallen dramatically in the 1940s and '50s (119–20), and, by 1963, the U.S. Fish and Wildlife Service notes that there were only 417 known nesting pairs of Bald Eagles in the United States (U.S. Fish and Wildlife Service 2007). As a result, the U.S. Environmental Protection Agency banned the use of DDT in the United States in 1972 (Environmental Protection Agency 1972), aided in part by the passage of the Federal Environmental Pesticide Control Act of that year (Environmental Protection Agency 1975). In the early 1970s, the number of nesting pairs of Bald Eagles was already rebounding due to the end of the Alaskan bounty and a decreased use of pesticides.

The importance of environmental laws and scientific research to the protection of Bald Eagles is only part of the Bald Eagle's story. The other part has to do with the growth of strong feelings of environmental nationalism around the species. Environmental nationalism, in the case of the Bald Eagle, is a cultural formation that has taken the symbol of the Bald Eagle as an exceptional, national bird and intensified that iconography, transforming the species into a patriotic figure. This transformation has taken place in part because the eagle is now associated with what is seen as an environmental success story. Said another way, the existing symbol of "national bird" has been combined with the understanding of the Bald Eagle as environmental success story to create a patriotic environmental image. That is how environmental nationalism has worked itself out in relation to the eagle, and one cannot fully understand the position of the Bald Eagle in U.S. environmental law or sentiment without understanding this ideological component of how the Bald Eagle is perceived and treated.

In a range of disciplines, scholars have discussed the idea of environmental nationalism, applying the concept in various ways. By and large, the term has been used to make sense of how large-scale environmental issues pertain to international politics and more localized formations of nationalism. For instance, Andrew Hurrell discusses how environmental nationalism relates to Brazilian economic policy in terms of the pressing issue of Amazonian deforestation (1993), and Reuel Hanks describes the role of environmental nationalism in response to the collapse of the Aral Sea and resulting formation of new national identities in Uzbekistan and Kazakhstan (2000). Somewhat differently, Heather Goodall describes how groups of settlers have created national origin stories and myths about themselves that involve heroic environmental encounters, even if those narratives ignore environmental effects (Washington, Goodall, and Rosier 2006, 88). How this concept of environmental nationalism can be put to work to help understand what has happened in the case of the Bald Eagle is somewhat different.

As mentioned, the Bald Eagle has long been a symbol of the nation in U.S. iconography (Lubbers 2000, 18), but belief in the Endangered Species Act as redemptive and successful, in the case of the Bald Eagle, has spawned environmental nationalism around the bird. The Bald Eagle, enmeshed as it now is in this aura of environmental nationalism, is seen as triumphant in the face of environmental degradation, and the nation by relation becomes constructed as able to repair itself and recover from environmental mistakes. We see this very clearly in Alaska, the place where Bald Eagle talons were once gathered in 55-gallon drums.

Today, the proliferation of living Bald Eagles in Alaska is billed as a key draw for ecotourism there. Birdwatchers can see large numbers of Bald Eagles feeding on spawning salmon in many locations, and events such as the annual Alaska Bald Eagle Festival in Haines draw in birdwatchers and other ecotourists alike. Tourists in Alaska pay to take Bald Eagle tours, and the Bald Eagle is used in the marketing of Alaska as a tourist destination in ways that brand Alaska, the state that is otherwise disconnected from the contiguous states and just miles from Russia, as particularly American. Images of the eagle are used to conjure up a sense of resilience, strength, recovery, and health for the state, bird, and country. Outside of Alaska and across the United States, Bald Eagles are living environmental symbols of a vibrant, strong American nation.

That is what Mark Barrow refers to when he calls the Bald Eagle a "quintessential example of charismatic megafauna" (Barrow 2002). Since the 1970s, the Bald Eagle has been reemerging as one of the country's most resonate national symbols (along with the American flag) in what is now a post-9/11 American patriotic culture (Skitka 2005). In sports stadiums in the United States, a Bald Eagle named Challenger has been trained to circle playing fields before landing on the gloved hand of its trainer. Challenger, named to memorialize the NASA space shuttle that exploded in 1986, is managed by one of several pro-eagle patriotic groups and foundations, the American Eagle Foundation (www.eagles.org). Challenger excites sports fans before the singing of the national anthem and during halftime shows as a kind of living icon of environmental nationalistic spirit.

Whereas the Bald Eagle was once considered a bad citizen and hunted as a bounty bird, they are now unequivocally described as "majestic" and promoted as unwavering symbols of the nation. In the case of the Bald Eagle, science, law, and sentiment all come together to inform how this bird is treated, in terms of environmental management, and considered, in terms of our popular environmental imagination.

So, where does this leave birdwatching in relation to the Bald Eagle? Describing environmental activism leading up to the passage of the Bald Eagle Protection Act in 1940, Barrow explains the contributions that birdwatchers made to raptor preservation on the whole in North America:

> The ongoing Audubon campaign was only one of several factors that led to heightened protection for the eagle and other American birds of prey. Perhaps more critical was the expanding interest in these species that accompanied the growth of birdwatching as a mass leisure activity. . . . A strong devotion to birdwatching was the common denominator that linked virtually every early bird-of-prey advocate in the 1920s and 1930s. In the process of regularly viewing raptors in the wild, these individuals developed a strong emotional and aesthetic bond with predatory birds, a bond that led them to speak out on their behalf. As birdwatching took hold, increasing numbers of other Americans sought to view birds of prey in their native environments. (Barrow 2002, 87)

As mentioned, since the publication of Peterson's popular guide in 1934, Bald Eagles have been described in guides with increasing detail and yet without mention of their changing status and treatment. Field guides teach birdwatchers such things as how to differentiate immature Bald

Eagles from similar-looking Golden Eagles, and an increasing number of facts about vocalizations and habitat have been included in each successive field guide. What technical, Petersonian field guides have not done, however, is encourage birdwatchers to see Bald Eagles as deeply embroiled in legal, scientific, nationalistic, and sentimental debates over their near extirpation and ultimate protection.

To singularly adopt the perspective of technical field guides would mean to fail to register all that is associated with birds, but many birdwatchers see and relate to birds in ways that exceed this kind of single-mindedness. As Barrow describes, for instance, birdwatchers came to appreciate raptors before environmental protection laws did, which made a difference in terms of helping sponsor avian protection. Technical field-guide literature focuses on what is necessary to identify a species, though, and it comes at the expense of what is needed to develop a richer, more complex knowledge of birds. In the case of the Bald Eagle, this division has not ended up having lasting negative consequences; recovery efforts have been successful. But the success story of the Bald Eagle, which involves both the bird's recovery and the removal of the "nuisance bird" label, is far from the norm.

MUTE SWANS: THE UGLY DUCKLINGS OF BIODIVERSITY

Hans Christian Andersen's story "The Ugly Duckling," first published in 1843, hinges on the belief that mature swans are among the most gorgeous birds in the world. Although the ugly duckling suffers as an "ugly" juvenile bird, when it matures it joins a beautiful flock of adult swans. Pyotr Ilich Tchaikovsky's ballet *Swan Lake*, written in the 1870s, is similarly based on the belief that swans are elegant, gracious birds. Would that ballet be the same had it have been called *Pigeon Lake*? What about *Gull Lake*? The simple answer is no, because particular species of birds have long been evocative of strong associations in Western thought and culture. Given their legacy, Mute Swans still conjure up images of beauty, elegance, and grace.

Although "The Ugly Duckling" and *Swan Lake* are European and Russian stories, respectively, love for swans in the United States in the nineteenth century was no less developed; significant numbers of Mute

Swans, which are native to central Eurasia, were imported to North America during this period in attempts to beautify the lakes of estates and parks.[7] As is often the case with imported birds of this kind, the Mute Swans did not remain on the ponds they were intended for, and at present the Cornell Lab of Ornithology's website notes that "escaped individuals have established breeding populations in several areas, where their aggressive behavior threatens native waterfowl" (2003).

Since roughly the 1960s, substantial populations of Mute Swans have taken hold in most eastern states, with particularly large numbers inhabiting Chesapeake Bay (Perry, Osenton, and Lohnes 2001). There, Mute Swans have been known to put pressure on other species, prey on some birds, and consume large amounts of aquatic vegetation. There is a clash, then, between a long-standing appreciation of swans and the Mute Swan's relatively newfound status in North America as a nuisance bird. Mute Swans are in the midst of legal and scientific debates, even as many birdwatchers and other bird lovers continue to value and appreciate them.

As in the case of the Bald Eagle, law provides one perspective on this issue. The Migratory Bird Treaty Act of 1918 offers broad protections for many species of birds, making it unlawful to

> pursue, hunt, take, capture, kill, attempt to take, capture or kill, possess, offer for sale, sell, offer to purchase, purchase, deliver for shipment, ship, cause to be shipped, deliver for transportation, transport, cause to be transported, carry, or cause to be carried by any means whatever, receive for shipment, transportation or carriage, or export, at any time, or in any manner, any migratory bird, included in the terms of this Convention . . . for the protection of migratory birds . . . or any part, nest, or egg of any such bird. (16 U.S.C. 703)

Key language in this law is the phrase "any migratory bird, *included in the terms of this Convention*" (emphasis added) because this clause renders species not listed in the law without coverage.

Mute Swans were originally one of several species left unprotected under the Migratory Bird Treaty Act of 1918; as a result, the species was aggressively managed when populations began to spike in the 1960s and '70s. In the states with Mute Swan populations in the thousands of birds, such as Rhode Island and New York, legal management methods have included authorizing landowners to hunt the swans, large-scale egg addling,[8] the destruction of nests, hazing birds, and culling large flocks (Atlantic Flyway Council 2003). A 2001 court order abruptly reinterpreted the Migratory Bird Treaty Act of 1918, offering new protections for Mute Swans, but this decision was soon voided by passage of a new law, the Migratory Treaty Reform Act of 2004 (U.S. Fish and Wildlife Service 2006). The Reform Act of 2004 alters the original Migratory Bird Treaty Act of 1918 by making it applicable only to species native to North America (Congressional Budget Office 2004), thus delisting nonnative Mute Swans once again from legal protection.

In the Chesapeake Bay area, where Mute Swan populations are particularly abundant, there were an estimated 3,000 to 4,000 Mute Swans in 2002. Acting with the freedoms allowed by the Reform Act of 2004, Maryland's Department of Natural Resource Management is reported to have killed 2,700 Mute Swans between 2005 and 2006 and more Mute Swans in 2007 (Humane Society 2008). It is through legal mechanisms, then, that Mute Swans have been actively controlled by a range of state and federal agencies committed to keeping their numbers in check. In this case, environmental law has enabled the aggressive management of the species, and the main reason is because of arguments made by biologists that Mute Swans threaten biodiversity.

In that Mute Swans are nonnative to North America and known to alter native landscapes, they are members of a much larger class of plants, animals, and insects labeled "invasive" in North America and seen as needing aggressive management as a result. In a recent summary of the management of invasive vertebrates in the United States, authors Gary Witmer, Patrick Burke, Will Pitt, and Michael Avery list "94 species of introduced/invasive birds [that] have become established in the U.S.," stating that "most introductions were as pets, but many were introduced for sport hunting" (2007, 128). Of note in this report is the slash between "introduced" and "invasive," a marker indicating how most species that are now considered invasive did not actually "invade" North America as get introduced by humans. Of the ninety-four species of birds currently considered nonnative to and thus invasive in North America, many were introduced and others are populations that resulted from escaped pets. The vast majority of listed species are construed as a nuisance because they pose some threat to biodiversity.

So, although federal law allows for aggressive management of the Mute

Swan and although environmental protection agencies have argued that the presence of the Mute Swan is a threat to existing forms of biodiversity, planned killings of the birds around Chesapeake Bay in 2003 and subsequent killings under the Reform Act of 2004 have sparked public outrage. Animal rights groups have opposed the legally permitted killings, with the Humane Society publishing articles labeling the Reform Act of 2004 excessively "nativist" in protecting only indigenous species (Markarian 2005). Gary Alan Fine and Lazaros Christoforides have described the law as an example of "biological nativism" involving "hostility to these birds who found American soil so much to their liking" (1991, 377).

In published appeals for Mute Swan preservation, the Humane Society goes back to the trope of the Mute Swan's beauty. For instance, in a FAQ published in 2008, the animal rights group argues that "from a macro perspective, the relatively recent arrival of a few thousand mute swans is just a minor—and incredibly beautiful—variation in the annual species composition of waterfowl in the bay." And later in that same document, it is argued that "the DNR's [Department of Natural Resources] lethal management plan arbitrarily targets a beautiful element of the Chesapeake Bay ecosystem, causing the unnecessary and inhumane killing of these animals, reportedly using cruel methods" (Humane Society 2008). Beauty is still assumed as an inherent quality of this nuisance bird.

With animal rights groups and bird lovers opposing the killing of Mute Swans, public messages about the Mute Swan have been mixed. In 1997, seven years before the DNR in Maryland ramped up its efforts to create "swan-free areas" in Maryland (Maryland Department of Natural Resources 2003), the U.S. Postal Service released its new "Love Swans" stamp, featuring two Mute Swans as its central image. On the stamp, the two swans form the shape of a heart with their heads and necks, creating associations of love and beauty with the Mute Swan. The stamp could have been printed in the 1870s, when *Swan Lake* was first being performed, as it embodies the long-standing perspective that swans are intrinsically beautiful, special, and graceful; because Mute Swans usually mate for life, they have long been used as avian representatives of Western romantic ideals.

So, while the U.S. Fish and Wildlife Service was working to limit the populations of Mute Swans, the U.S. Postal Service was helping sustain a belief in the Mute Swan as a special bird. Bald Eagles, as shown, ultimately gained both legal and cultural support, but the story of the Mute Swan is different. Mute Swans had long had cultural support but were recently cast by government agencies as a threat to biodiversity. Much as killing a "bird of prey" listed as a bounty bird was once seen as a form of justice, since the 1990s killing Mute Swans has been constructed as such. In a white paper on the topic of the Mute Swan, the Pennsylvania Game Commission refers to the vital role that public opinion plays in the management of birds. The white paper reads, "Regardless of the exact control methods ultimately adopted [to decrease numbers of Mute Swans], it is very important to change public perception that mute swans are a wonderful addition to our outdoors" (Gregg 2005). In avian environmental management, how birds are perceived is significant.

Birdwatchers have their own and quite different investments in Mute Swans. Swans are part of a group of species that are particularly challenging to distinguish from one another in the field because the visible and audible differences between the four species of swans found in North America (Mute Swans, Trumpeter Swans, Tundra Swans, and less frequently Whooper Swans) are so subtle. Mute Swans have orange on their bills, but juvenile Tundra Swans have pinkish bills that can appear orange. All four species are white when mature, but Mute Swans sometimes raise their wings a bit while swimming and hold their necks in more of an S-shape. Trumpeter Swans can be distinguished from the others because their necks are somewhat straight up and down, and all the swans have slightly different calls (Allen and Hottenstein 2002, 66–67). At places where Mute Swans are found, such as the Chesapeake Bay area, viewing platforms make it possible for birdwatchers to see and identify the birds.

It is certainly not true that Mute Swans are the most sought-out birds in North America, but they do provide a draw for birdwatchers interested in making complex identifications. If a birdwatcher is having trouble identifying a swan, any contemporary field guide can be of help; Mute Swans have appeared in every guide to North American birds since Roger Tory Peterson's 1934 *A Field Guide to the Birds.* A few recently published field guides allude to a contentious debate surrounding the Mute Swan (Allen and Hottenstein 1983, 66; Kaufman 2000, 42), but David Allen Sibley's *The Sibley Guide to Birds* (2000) makes no mention of this issue.

The Mute Swan, a species that has long been revered, is in the process of being recast as a nuisance bird in need of aggressive management.

Field guides published since the 1990s, since efforts to decrease Mute Swan populations have been stepped up, encourage birdwatchers to identify this species of bird even while it is the target of systematic removal. While environmental managers work to extirpate Mute Swans and animal rights activists work to save adult Mute Swans, birdwatchers go out in search of this hard-to-identify bird. There is no doubt that the ways Mute Swans are valued in field guides is quite different from the romanticized view of swans captured in the U.S. Postal Service's "Love Swans" stamp, but it seems worth asking what side (if any) of this debate field guides are on. Does representing Mute Swans in taxonomic ways in the pages of a field guide amount to a complicit endorsement of anti–Mute Swan measures, or is it a disavowal of them?

For the birdwatcher, the Mute Swan has value as a North American bird that can potentially be added to his or her list of identified species. The Mute Swan also provides a good opportunity for practicing the skills of species identification. This value is not diminished, however, by the reduction of Mute Swan populations. Even if populations of Mute Swans were cut by 90 percent in North America, there would be plenty of Mute Swans for birdwatchers to identify. In fact, if the numbers of Mute Swans decreased that much, scarcity would make finding one as exhilarating as finding an endangered Bald Eagle once was. Birdwatching as an environmental practice, then, fits as easily with the lethal forms of management being meted out against Mute Swans as it does with the preservation of such species as the Bald Eagle.

It is preservation, though, and not the aggressive management of nuisance birds, that is most often cited as fundamental to birdwatching. The notion that saving birds is in the best interest of birdwatchers helps construct birdwatching as a green pastime even though, in fact, killing birds serves the interests of birdwatchers, too.

EXPENDABLE GULLS

The nuisance birds discussed so far fall into two categories. The Bald Eagle is a native bird that was seen as a nuisance but has since been protected by new laws and changing perceptions, and the Mute Swan is a nonnative bird that is in the midst of being recast as a nuisance because it threatens biodiversity. Several gull species fall somewhere in between by being native to North America and yet still seen as nuisance birds. All these birds—Bald Eagles, Mute Swans, and gulls—are of great interest to many birdwatchers. Gulls are like swans in being intriguing because they are so difficult to differentiate.

In recent history, many large colonies of Herring Gulls have been culled in the United States and Canada (Duncan 1978; Kress 1983; Guillemette and Brousseau 2001); such killings have frequently been at airports, given the danger large flocks of birds pose to jets (Smith and Carlisle 1993). Although U.S. law protects gulls, such laws also allow for permitted killings. In this section, I look at a particular culling of gulls at Monomoy National Wildlife Refuge on Cape Cod, a birdwatching hotspot on the eastern migratory flyway. This management initiative was implemented with the goal of protecting the habitats of two rarer bird species, but it turned into a management fiasco when dead and dying gulls started washing up on the beaches of Cape Cod.

In 1996, the U.S. Fish and Wildlife Service poisoned a large gull colony consisting largely of Herring Gulls and Great Black-backed Gulls at Monomoy (Schoch and McKinney 1996, 69). The federal agency used bread laced with the poison DRC 1339, intending to kill the gulls in their nests, and ultimately exterminated somewhere between two thousand and six thousand birds resident at the site (Rimer 1996; Melley 1997). Gulls at the refuge were targeted because they were seen as encroaching on areas set aside for populations of two other bird species: endangered Piping Plovers and Roseate Terns. To create what it called a "'safe harbor' for coastal birds," the U.S. Fish and Wildlife Service worked to establish a "'gull-free zone' on a portion of South Monomoy" (U.S. Fish and Wildlife Service 2001, 7). This wording is somewhat contradictory, as gulls are coastal birds too, but no matter: the gulls were seen as nuisance birds by nature of being both superabundant and inclined toward preying on the tiny Piping Plover chicks.

As birds that are labeled "endangered," the plovers and terns are protected by the Endangered Species Act of 1973, which does not protect Herring Gulls and Black-backed Gulls. The more powerful 1973 Endangered Species Act allows for a sort of override of the Migratory Bird Treaty Act of 1918. Recall from the previous discussion of the Bald Eagle that the Endangered Species Act can be used to preserve and protect the *habitats* of endangered species, not just the species themselves; thus, thousands of the more abundant gulls were killed in 1996 and subsequently managed on the site on an ongoing basis to protect the plover and tern habitat.[9]

The killing of the Monomoy gulls was supported by a strong investment in a kind of utilitarian equation in which killing a large number of birds from a "superabundant" group of nuisance birds is seen as justifiable to make room for a small number of birds from a lagging species. Two thousand common birds killed for ten or twenty rare birds saved makes sense in this logic. Numbers of individual birds are seen as less valuable than the total number of species. Biodiversity trumps biomass. Note that in arguments to save Mute Swans the Humane Society relied on descriptions of swans as "incredibly beautiful," but similar claims were not made in support of the gulls at Monomoy. Gulls, like other abundant scavengers, are often imagined as plain or even filthy.

Although books such as Richard Bach's *Jonathan Livingston Seagull* portray gulls in a positive light, by and large our popular environmental imagination does not favor gulls. Several animal rights groups opposed the killings at Monomoy, but public outcry was focused not on the loss of the gulls but on the inhumane use of DRC 1339 to kill them. (The poison worked slowly on the gulls, giving them time to fly to more populated areas of Cape Cod before dying slowly.) Representatives from the Humane Society worked to humanely euthanize suffering gulls as they washed up on nearby beaches (Rimer 1996), local and national newspapers printed the story, and the culling at Monomoy became an environmental management debacle.

At Monomoy, the U.S. Fish and Wildlife Service saw the gulls as an expendable nuisance that impinged on the viability of the more valuable Roseate Terns and Piping Plovers, and animal rights groups saw the poisoned gulls as victims of animal cruelty. Many birdwatchers tend to find value in gulls because they form a particularly challenging set of birds to identify. Although not all birdwatchers are particularly interested in gulls, many avid birdwatchers are keen on them for two reasons: the nearly thirty species of gull in North America are difficult to differentiate in the field, and immature gulls can be positively baffling to distinguish because gulls begin their lives as brownish-grayish birds and only slowly, over several years, develop distinguishing adult plumage.

Although many nonbirdwatchers see "seagulls" (a misnomer) as all alike, avid birdwatchers are known to relish in the identification challenges posed by mature and immature representatives of different gull species. Field-guide author Kenn Kaufman captures this sentiment as well when he writes: "Much more striking is the variation in young birds. Small gull species may reach adult plumage in only two years, but larger ones take three or four years, their plumage patterns changing throughout this time. A flock of gulls can show a dizzying array of different patterns, even if the birds all belong to the same species!" (2000, 68).

Just because gulls are of interest to many avid birdwatchers, though, does not mean that the birdwatching community has rallied in opposition to cullings of the kind seen at Monomoy. Although exterminations of this kind might upset some birdwatchers, lethal avian management is widespread. At the many airports where gull colonies have been culled in the interest of air travelers, there is no question that the birds have died to serve human interests. At places such as Monomoy National Wildlife Refuge, however, the culling of thousands of gulls has been billed as a management decision made exclusively in the interest of plovers and terns.

Reducing numbers of superabundant species runs parallel to at least some of the interests of birdwatchers. First, such control measures are intended to save the rare species birdwatchers seek, in this case Roseate Terns and Piping Plovers. In this sense, avian environmental management and birdwatching share a certain calculus: in both arenas, large numbers of a single superabundant species have less value than small numbers of an endangered one. Second, eliminating large flocks of Herring and Black-backed Gulls makes them locally scarcer. Avian scarcity is a condition in which birdwatchers thrive, prompting assertive birdwatchers to drive or fly many miles in search of a single rare bird. That is certainly not to say that the birdwatchers around Monomoy rejoiced at the thought of having to search for common species of gulls, but rather to suggest that birdwatching thrives under the simultaneous conditions of scarcity and biodiversity.

Wildlife refuges like the one at Monomoy are prime birdwatching destinations and are maintained as such. In a brochure about the area, the U.S. Fish and Wildlife Service describes plovers as particularly worth seeking out because they "have been described as everything from wind-up toys to cotton balls with legs rolling along the sandy beaches of the coast" (U.S. Fish and Wildlife Service 1997). At Monomoy, when birdwatchers aim their binoculars at plovers or terns, what they see is in part made possible by the locally reduced numbers of gulls.

In the case of crows, the next and final nuisance bird to be discussed, a different situation is at hand. Here, hunters have been enlisted to manage multiple crow species as if all crows are alike, something avid birdwatchers know all too well not to be the case.

THE INDISTINGUISHABLE CROW

Crows are similar to gulls in that they are native to North America, thought by many to consist of just one species, and have a long-standing reputation as nuisance birds. In the 1890s, when Mabel Osgood Wright was referring to crows as cowards with "a hoarse voice and disagreeable manners" (1895, 179), ornithologists had yet to determine that there are four separate species of crow commonly found in North America: the American Crow, Northwestern Crow, Tamaulipas Crow (formerly called the Mexican Crow), and Fish Crow.

To the untrained eye, each species looks and sounds generally alike, but to birdwatchers who know their crows, the different species can be distinguished via bundles of subtle field marks, including size, shape, vocalization, coloration, and regional distribution. David Allen Sibley, in his *Sibley Guide to Birds* from 2000, concludes that "crows are best identified by voice" (360) because the other field marks are so difficult to discern. Because crows comprise separate species that are hard to distinguish in the field, they are akin to swans and gulls in providing avid birdwatchers with particularly satisfying identification challenges. At the same time, some birdwatchers are as uninterested in crows as they are in gulls, in part because neither group of birds is particularly rare or seen as exotic.

Being largely indistinguishable as different species poses problems for crows in U.S. environmental law. "Crows" on the whole were not initially granted protection under the Migratory Bird Treaty Act of 1918, and although the 2001 legal review of that act granted crows protected status, such protections have several limitations. As the above discussion of gulls has already shown, permits can be obtained from the federal government to allow for aggressive management, and federal law also allows states to sponsor annual crow-hunting seasons. Furthermore, because crows are known to damage agricultural interests, landowners can kill crows without a permit if the birds are "found committing or about to commit depredations upon ornamental or shade trees, agricultural crops, livestock, or wildlife, or when concentrated in such numbers and manner to constitute a health hazard or other nuisance" (New York State Wildlife Control, n.d.). This extremely open language makes it legal to kill crows under almost any circumstance. None of this legislation attends to the different species of crows found in North America, but instead addresses all crows as if they were a single species.

The American Crow is the species of crow that is most widespread in North America, and this is the species that is also most known to damage crops, but because laws do not distinguish among the four different species of crow, all crows in North America are managed as if they were American Crows. Although crows are no longer actively hunted as bounty birds in most U.S. states,[10] current hunting laws are extremely lax in having no bag limits. In addition, although the federal government has determined a 124-day hunting season for crows, as Kevin J. McGowan notes, several states spread crow-hunting season out over eight months by limiting crow hunting to certain days of the week (McGowan 2001). With no bag limit and extended seasons, hunters can kill dozens and even hundreds of crows in a single day, even though crows are seldom eaten for food. Pro-crow-hunting groups such as Crow Busters (www.crowbusters.com) take an aggressive outlook toward crow hunting, presenting it as a sport that provides service to the environment in minimizing damage done by "pests."

One of the less widespread species of North American crow is the Fish Crow, and although it is found as far west as Ithaca, New York, it is most common in the eastern and southeastern United States. For even the avid birdwatcher, distinguishing between a Fish Crow and an American Crow can be a real challenge: all the differentiating field marks are subtle and hard to detect in the field. The two species have different calls, for instance, although the call of the Fish Crow can be mistaken for the begging sounds of the American Crow. Fish Crows are generally smaller than American Crows, but size variation among American Crows means that some of them may be smaller than Fish Crows (Cornell Lab of Ornithology 2002). Telling crows apart has thus become a minor hobby for some birdwatchers.

In Virginia, a state where the habitats of Fish Crows and American Crows overlap, crow-hunting season runs for seven months, from "August 16–March 21 on Monday, Wednesday, Friday, and Saturday only," and the species of crow that hunters can kill is not specified (Virginia Department of Game and Inland Fisheries 2008). Although avid birdwatchers in Virginia work to tell species of crow apart, Fish Crows and American Crows are seen as interchangeable by this legislation and the hunters who operate within it. As the state's Department of Game and Inland Fisheries explains on its website, "Crows are a federally regulated migratory species (sic); however, no HIP number [Harvest Information Program, or hunting permit] is required to hunt them" (2008).

A blatant disregard for speciation among crows takes place in the hunting laws in Washington State as well, where another species of crow, the Northwestern Crow, sometimes mixes in with populations of American Crows. The Washington Department of Fish and Wildlife makes no distinction between these two species in its regulation of Washington's crow-hunting season, even though avid birdwatchers in the state puzzle over their crow identifications. Hunting season for "crows" in Washington runs from October 1 to January 31, and it is noted somewhat inaccurately on the state regulatory agency's website that "crows in the act of depredation may be taken at any time" (Washington Department of Fish and Wildlife 2008). Although most other hunted animals have a bag limit in Washington State, including all other species of hunted birds, hunters may kill as many crows as they like.

The four species of crow regularly found in North America are in an unusual position. The widespread American Crow has garnered a reputation for all other species of crow as a nuisance, in part because most people do not know about or recognize biodiversity among crows. As a result, American Crows, Fish Crows, Northwestern Crows, and Tamaulipas Crows of south Texas[11] can all be hunted, regardless of the populations and behavioral particularities of each species.

All this crow hunting might not be a problem from the point of view of species viability if not for the recent spread, starting in 1999, of West Nile virus among crows and other corvids. One recent study of crow populations in North America shows that, since the outbreak of West Nile virus, American Crow populations have been hardest hit; by 2005, numbers of American Crows were down by as much as 45 percent from 1998 levels (LaDeau, Kilpatrick, and Marra 2007, 710). This study, published in the journal *Nature*, begins by stating that "emerging infectious diseases present a formidable challenge to the conservation of native species in the twenty-first century" (710). This point is worth noting because the authors of this article clearly see American Crows as a threatened native species. Indiscriminate hunting laws meant to manage crows do not see crows that way and compound the threat of West Nile virus.

Crows are under pressure from multiple directions, and in the new epidemiological context of West Nile virus, crows are being recast by some as a threatened natural resource. Although protecting American Crows has never been a concern of U.S. environmental law, now, with the spread of West Nile virus, some advocates for the widely hunted "crow" are beginning to emerge. Recent news stories have been published in the popular press, for instance, lamenting losses of crows and describing crows infected with West Nile virus as equally valuable as bluebirds and robins (Minnesota Public Radio 2007). A new form of sympathy may be emerging for crows, as research shows them to be intelligent toolmakers (Savage 2007), and if the experiences of the Bald Eagle, Mute Swan, and Herring Gull are any indication, sympathy will be needed to reverse the current management practices that see all crows as an indistinguishable nuisance.

WHY NUISANCE BIRDS MATTER

This chapter began with a description of how, as field guides changed in the 1920s and '30s, strongly judgmental statements about individual species of birds were replaced by a linguistic and visual system of representation that focused much more strictly on field marks and taxonomy. Field guides became technical tools intended to help birdwatchers identify birds, not prefer one species over another, and as such they produce an appearance of having to become divested of environmental issues and debates. Of course, this seemingly unbiased system of representation is filled with another set of biases, such as an investment in scientific knowledge and a tendency to view birds as easily abstractable from environmental and cultural contexts.

But birds do not exist apart from their immediate habitats, just as they do not exist apart from cultural beliefs and environmental laws. Our perceptions of certain species and groups of species have serious consequences in terms of how birds are treated and managed. More than one hundred thousand Bald Eagles were killed in Alaska for a bounty, and thousands of Mute Swans and gulls have been culled. Multiple species of crows are currently being hunted without bag limits in many states, even as American Crows face new pressures from West Nile virus. The nuisance birds discussed here are only a small sampling of the nearly one hundred species of birds that are currently defined as a nuisance in North America. Most have gained this designation because they are nonnative, but native species of crows and gulls are also on the list for a range of reasons, both real and imagined.

A species' status often changes over time and can vary by region. Discussing the sometimes contradictory laws related to grackle conservation and extermination, Dorie Bargmann writes: "In 1989, a federal court in Illinois found Henry Van Fossan guilty of poisoning two mourning doves and two common grackles with strychnine. He was fined $450 and given three years' probation for violating the Migratory Species Protection Act. In 1992, the federal government authorized the killing of 164,478 grackles nationwide. In Texas alone, in fiscal year 2000, 17,095 great-tailed grackles were poisoned under the auspices of the federal Wildlife Services program" (2005, 139). Except in cases in which a bounty has been placed on certain birds, killing nuisance birds without a permit can be illegal. Permitted killings, however, are widespread, aggressive, and frequently funded by taxpayer dollars.

In some of the cases discussed, environmental laws relating to nuisance birds are misinformed, but what is at issue in this chapter is less whether existing environmental laws and management practices regarding nuisance birds are good, ethical, or productive. Instead, the point has been to show that although birdwatching is typically represented as a practice informed singularly by a "love for birds" that then leads to avian conservation, love and conservation are only part of the picture. Aggressive forms of avian management continue to rely on much darker sentiments and a belief that nuisance birds can be empirically determined.

Technical field guides ignore much of that and instead represent a seductively simple vision of the world birds live in: according to technical field guides, all birds are equal and able to live out untroubled lives in untrammeled landscapes. Such is the message of one of the central tools birdwatchers use.

Published histories of birdwatching have tended to describe the many contributions birdwatchers and birdwatching have made to bird conservation in particular and conservation more broadly. Capturing this view, historian Mark Barrow describes how, in the twentieth century, decades of campaigning on behalf of "a network of bird enthusiasts" helped repair the reputations of raptors in ways that ultimately enabled their protection (2002). Birdwatching, when understood through this lens, appears to have facilitated not only the protection of birds, but the development of conservationism in North America. Another way of seeing birdwatching is as hands-off and thus harmless to the birds birdwatchers spend their time looking at (Barrow 2002, 115).

It is worthwhile to complicate both of these views of birdwatching while not erasing that birdwatching and birdwatchers have indeed been integral to environmental conservation. Paying attention to nuisance birds matters because antipathies toward them are indicators of the ongoing role of sentiment in environmental management. Their often lethal management is a reminder that "saving birds" has never been an issue of protecting absolute numbers of birds, but merely numbers of species.

The next chapter moves to examples of what a more richly contextualized view of birds can look like. By focusing on representations of birds in modified environments, birds can be used to signify competing beliefs in environmental decay and resilience.

CHAPTER THREE

Picturing Birds in Altered Landscapes

The main purpose of field guides is to help birdwatchers identify birds. As I have been arguing all along, though, these books have always done much more than that. The authors of the first birdwatching field guides used extended narratives to create new feelings about birds that, in turn, were intended to fuel conservation. Blueprints for this kind of emotional birdwatching became less evident in technical field guides that, through their language and artistry, promote a vision of birds as divorced from cultural and environmental concerns.

One such environmental concern is toxic pollution. In some cases, birds by the thousands have been killed by environmental pollutants (Patterson 1999); other bird populations suffer from the ill effects of having high levels of toxic substances, including heavy metals, in their bodies (Roberts 1997). Although pollution is a pressing environmental problem affecting not only birds, the mainstream field guides that birdwatchers use skirt the issue, representing birds as living out healthy lives in clean environments. By failing to reshape the conventions of scientific bird illustration to engage with contemporary environmental issues such as pollution and toxicity, field guides sanitize the landscapes that birds live in. This hygienic system of representation encourages birdwatchers to look for birds while explicitly not looking for toxic waste and the host of other environmental problems toxicity represents.

There is one powerful exception to the disavowal of toxic pollution in contemporary field guides, however. Jack Griggs's *All the Birds of North America* (1997) is unique among contemporary field guides in that several of the images in Griggs's guide present birds living in landscapes populated with human technologies, pollution, and even birdwatchers themselves. This chapter centers on Griggs's atypical representations of birds in modified landscapes. *All the Birds of North America* positions birds in polluted and populated landscapes, but those images result in a good deal of ambivalence in terms of their environmental message, alternatively suggesting that birds are either suffering in the midst of an environmental crisis or thriving despite human encroachment.

To explore the range of readings that come out of Griggs's altered landscapes, I turn to similar representations of birds from two very different contexts: contemporary advertising and art. Taxonomic representations of birds in different contexts show how fastidiously accurate representations of birds get used to further a range of arguments about birds, technology, pollution, and "nature" as a category. One set of images I explore comes from a series of bird portraits by John James Audubon that were modified in a General Electric Company (GE) ad campaign. Although GE's ads promote the idea that birds thrive in altered landscapes, several paintings by the contemporary environmental artist Alexis Rockman show birds suffering due to human technology and pollution. Rockman's paintings suggest not only that birds suffer due to their encounters with humans, but that their recognizability, and thus taxonomic status, can be compromised as well.

By turning to these three sources of visual material—Griggs's field guide, the altered Audubons from GE, and Alexis Rockman's art—this chapter engages in what Carey Jewitt and Rumiko Oyama have described as a social semiotic approach to visual analysis (2001). In this approach to reading visual images and culture, meaning is seen as "established by the syntactic relations between the people, places and things depicted in images. These meanings are described as not only representational, but also interactional (images do things to or for the viewer)" (Jewitt and Van Leeuwen 2001, 3).

In my readings of Griggs, GE, and Rockman, I attend in particular to visual features that create links to field-guide images. All three sets of images rely on the accuracy associated with field-guide imagery to assert very different claims about the environment. Such things as what is "natural" or "healthy" are inverted in some of the images to suggest that merely "watching" an environment, as bird*watching* suggests one do, amounts to an incomplete and in some cases flawed method of environmental interaction and assessment. Taking this negative evaluation of birdwatching one step further, in some of the images I discuss birdwatchers are included in scenes populated by birds to suggest some directly negative effects they can have on the birds they watch. Through a discussion of Griggs's field-guide images, GE's altered Audubons, and Alexis Rockman's dystopic environmental art, this chapter insists that there is a good deal at stake in how birds are represented and in the practices associated with observing them.

BIRDS AND TECHNOLOGY

In 1997, with sponsorship from the American Bird Conservancy, Jack Griggs published *All the Birds of North America*, a field guide infused with atypical environmental messages. Griggs intervenes in several established traditions upheld by more mainstream, technical field guides. Griggs, an accomplished digital illustrator, begins *All the Birds* with several haunting descriptions of extinct North American birds (the Great Auk, Passenger Pigeon, Ivory-billed Woodpecker, Labrador Duck, Carolina Parakeet, Heath Hen, and Bachman's Warbler). Whereas field guides tend to leave out extinct species, Griggs begins with them, creating entries that show how each recently extinguished species might have looked and acted when it was alive. Notably, he includes a discussion of how humans contributed to each extinction.

Although technical field guides are organized around taxonomic order, Griggs also disrupts this ordering by using his own system of organization based primarily on feeding behavior. By employing this alternate system of organization, Griggs asserts that knowing birds means knowing not just what they look like, but where they live and what they eat. Implied in this structure is the idea that habitats matter when it comes to knowing and preserving birds. Griggs's field guide includes several other reminders of the importance of responsible environmental stewardship, with a number of his images showing birds living near such things as buildings and train tracks.

Discussing Griggs's field guide, Raymond Korpi argues that, in choosing to represent birds in this way, Griggs goes against a long tradition in field-guide illustration that presents birds as separate from larger ecosystems. Korpi writes that the focus of field guides has traditionally been on a "single group of animals and on the specific task of identification, a private action unless the reader chose to take a proactive stance. But simply focusing on one aspect within an ecosystem went against the environmentalist's conception of the world" (Korpi 1999, 167–68). Although Korpi may oversimplify what an "environmentalist's conception of the world" entails, he is keen to point out that Griggs's field guide is radical in showing connections between birds and their complex environments. In Griggs's field guide, birds live amid a ubiquitous and sometimes troubling human presence.

The Horned Lark, for instance, forages alongside an active airport runway (figure 3.1).[1] In Griggs's representation of Horned Larks, the birds are in the foreground of a busy scene filled with several related species, and a jumbo jet is seen taking off or landing in the background. The Bay Bridge and a fog-shrouded San Francisco are visible behind the birds and the jet. In images such as this one in Griggs's field guide, "nature" is an inclusive realm including birds, plants, bodies of water, and such human contributions as airports, jets, bridges, cities, and trash. Birdwatching, according to Griggs, involves looking through binoculars to spot a bird's field marks but also lowering those binoculars to get to know the larger environments birds live in.

In terms of offering perspectives on the relationship among birds, field guides, technology, birdwatching, and environmentalism, figure 3.1 does several things. One is to argue that some bird habitats are under pressure from technologies such as airports and jet aircraft. In this reading of the image, the fragile birds can be seen to be eking out a living beside the unforgiving technology of the jet. An alternative way of reading Griggs's image, though, sees the birds as resilient in the face of technological encroachment. Interpreted in this way, the birds in figure 3.1 thrive despite their altered environment, suggesting that the birds may exist in harmony with the airport. These first two readings, one an environmental warning and the other an endorsement of current environmental practices, are hard to reconcile when looking at this image alone. In the larger context of the guide, however, the first reading (which sees

Figure 3.1. Sprague's Pipits, American Pipits, and Horned Larks (at right) pictured near the Oakland International Airport. Subtitle reads "East shore, San Francisco Bay." From Jack Griggs's *All the Birds of North America*, *1997: 120–21.*

the larks as imperiled) fits with Griggs's and the American Bird Conservancy's vision of where many of the birds of North America stand in terms of human alterations of the environment.

There is another way of thinking about the image, however, which I see as an even more significant function of images of this kind. To understand this, it is worth noting that the birds in figure 3.1 are rendered in much the same way as birds are depicted in every other contemporary technical field guide. The way the larks are posed and painted follows what Lorraine Daston and Peter Galison have described, in their work on nineteenth-century atlas makers, as "the precept of truth to nature" (1992, 84). Daston and Galison describe how nineteenth-century atlas makers in particular, but all kinds of scientific illustrators as well, aimed to represent "what truly is" (84) in their depictions of the natural world, and it is this illustrative principle that field-guide art draws on. That is not to say that field-guide illustrations are always entirely accurate, however.

Alan Gross has shown how, in some cases, ornithological knowledge and scientific depictions are created based on the rhetorical and visual oversimplification of small sample sets (1990, 33–53). Some of the work in ornithology that presents "what truly is," Gross demonstrates, is a kind of "rhetorical construction" that is part of "an interlocking set of persuasive structures" (34). So, although scientific illustration is based on precepts of being "true to nature," it is also, as Gross explains, prone to and part of a process of "tendentious simplification" (45). The images of birds in field guides unambiguously present images of birds as if they were visual replicas of living birds in nature. This imagery creates a sense that readers will be able to see the same things they find through the lens of their binoculars and spotting scopes as they find in their books. As Ann Shelby Blum describes in her discussion of nineteenth-century zoological illustration, "Acceptance of pictures as conveying authentic information about nature lies at the heart of scientific practice" (1993, 3).

The birds in Griggs's image are intended as just this kind of authentic and factually accurate map of living birds. As Daston and Galison remark: "Atlases supply working objects to the sciences of the eye. For initiates and neophytes alike, the atlas trains the eye to pick out certain kinds of objects as exemplary . . . and to regard them in a certain way" (1992, 85). Field guides do this same thing, and in many cases oversimplification helps birdwatchers identify the birds they seek. But what about the scenes birds are pictured in?

With the inclusion of additional elements such as airports, jets, bridges, and crowded cities in the background of an image centered on accurate (or seemingly accurate) depictions of birds, the added elements become relationally presented as part of "what truly is" because of a metonymic relationship between the accuracy associated with how the birds in Griggs's image are represented and the accuracy of everything else in the scene. Metonymic borrowing is created when two or more things, through association, gain meaning from one another, and that is how Griggs's very precise image of Horned Larks makes the presence of the jet aircraft, airstrip, bridge, and cityscape look as factual as the birds themselves. In the image, "authentic information about nature" includes not only the field marks of the different species of birds but the presence of human technologies in the lives of birds as well.

Griggs's field guide departs from other contemporary field guides to birds that create hygienic fantasies of what nature is "truly" like. *All the Birds* leverages how meaning is made in the tradition of bird illustration to claim that nature is thoroughly modified—and not necessarily for the better, when it comes to birds and birdwatchers. Human technologies are part of nature, and birdwatching should, according to Griggs, include looking more broadly at this larger technologized scene that many birds inhabit. Whether or not the jet, airstrip, bridge, and city are seen as encroaching on the birds in the image or leaving the birds unharmed is open to interpretation, but the presence of human technologies in the lives of birds is portrayed as incontrovertibly true given the logic of factual representation.

A strikingly similar image of birds and a jet aircraft, produced by GE, leverages Daston and Galison's "precept of truth to nature" much less ambiguously to deny that there are any negative environmental consequences of GE technologies. Before describing this ad in some detail, however, it is worth providing a bit of context about the series of GE ads this one is a part of. In 2005, GE launched what it calls its ecomagination advertising campaign. With ads in print, on television, and online, the ecomagination series creates a sense that GE's primary commitment is to producing such green commodities as fuel-efficient jet engines, wind and solar power, and desalinated drinking water.

To understand these ads, it is also important to note that they rely on playfully suggestive scenarios, so in one video ad, a tree walks across the countryside and hugs an energy-efficient house. In another video, fishermen pull plastic bottles filled with clean drinking water out of the ocean. In another ad, a group of half-naked models labor in a coal mine as a voice-over suggests that GE's coal-burning power plants are "clean." In one award-winning[2] subset of print ecomagination ads, GE's advertiser BBDO Worldwide hired illustrator John Perlock to digitally alter several images of birds created by John James Audubon to make a case for GE's products as green.

By using digitally altered Audubons, GE the polluter is nowhere to be seen, prompting at least one critic to describe the ad campaign as a case of corporate greenwashing (O'Donnell 2005). This strategy should come as no surprise. Founded by Thomas Edison in the 1890s (Bazerman 1999, 283), GE is one of the world's ten largest multinational corporations, with diverse holdings in a range of environmentally problematic industries, including energy and manufacturing. GE companies have contributed to multiple waste streams, dramatically altered landscapes, and caused toxic air and water pollution of various kinds.[3] In an attempt to revamp this image, GE and BBDO Worldwide turned (in part) to Audubon's birds.

Figure 3.2 is from 2005 and directly echoes Griggs's image of Horned Larks near San Francisco. The key difference is that, in GE's retouched Audubon, the jet aircraft flies by innocuously above and behind the birds. In addition to adding the jet to the image, GE has included promotional text at the top and bottom of the ad. In the case of the GE ad, there is an effort to unambiguously promote jets as unobtrusive, nonpolluting, and friendly to birds, which is in part accomplished by the placement of the plane farther back in the scene and away from the birds. It is also accomplished by the addition of an invented name for the species (*Lessus Pollutantus*) and text that reads: "The cleaner GEnx aircraft engine. Just one way ecomagination is creating a better world." The male Red-breasted Snipe (our Long-billed Dowitcher), pictured at the right in figure 3.2, points up to the plane using its outstretched beak, while the female bird (pictured at the left) forages peacefully at the water's edge.

Audubon's original painting of the Red-breasted Snipe has been changed in several ways. The original (figure 3.3) was based on a detailed study of the species that Audubon began along the Mississippi river in 1821 and continued through his travels for several years (Audubon 1999, 79–81, 150–51). In his original image, Audubon characteristically posed the birds for what Robert Welker has referred to as "dramatic effect." Audubon's original drama was about a scene on the ground, not in the air. Audubon was different from other bird illustrators before him in that he "animated poses and [created] exciting situations" (1955, 88); his images both identified species and dramatized their lives in real and imagined ways.

In the original image of the snipe, Audubon pairs male and female birds together to show their sexually dimorphic plumage while creating a chivalrous scene in which the female (at the left in figure 3.3) forages for food while the male stands guard with his long bill protectively behind and over her. Because Audubon represented birds dramatically and according to the conventions of precision and accuracy found in natural history illustration, and because he has since been made into a symbol of mainstream U.S. environmentalism by the formation of the National Audubon Society in the 1890s, GE latches onto his illustrations as a vehicle for its pseudo-green message. Audubon scholars tend

Figure 3.2. "*Lessus Pollutantus*," a GE ecomagination ad for the GEnx jet engine from 2005. Illustrated by John Perlock.

Figure 3.3 Audubon's original representation of the Red-breasted Snipe (our Long-billed Dowitcher). From *The Birds of America* by John James Audubon, 1840.

to agree that, by contemporary standards, John James Audubon was no environmentalist,[4] but GE appropriates his work because Audubon has been enshrined by the modern environmental movement. GE's advertisers have transformed Audubon's birds, as indicated by the humorous Latin coinage *Lessus Pollutantus*, if not Audubon's own personal philosophy about them.

It is curious, however, that GE would dare to represent its jet engines as "harmonious" with birds.[5] Although the GEnx engine may, as GE claims in one press release, be "the most fuel efficient, quiet, and low-emissions jet engine that GE has ever introduced for large jet aircraft" (General Electric Corporation 2005b), sharing the air with birds is not something jet engines do particularly well. In fact, the GEnx jet engine, like all jet engines, creates enormous amounts of suction and is a competent bird killer.

Nearly six thousand reported bird-strike incidents involving birds and civil aircraft have been reported each year in the United States since 2000, with bird strikes being the "second leading cause of aviation-related [human] fatalities" in the United States (Cleary and Dolbeer 2005, 251). An FAA report from 2008 states that "wildlife strikes have killed more than 219 people and destroyed over 200 aircraft since 1988" (Dolbeer and Wright 2008). To be certified for air travel in the United States by the Department of Transportation, the GEnx jet engine had to pass what is commonly referred to as the Department of Transportation's bird ingestion test. The test involves "passage of a bird into the rotating blades of a turbine engine," and a single five-and-a-half-pound bird and four two-and-a-half-pound birds (real or artificial) (U.S. Department of Transportation 2001, 1) were run through a GEnx engine before it could be certified at GE's testing facility in Peebles, Ohio.[6] Nonetheless, in the face of the deadly relationship between jet engines and birds and even though jet engines must be designed to ingest birds without failing, GE, in an act that could be called ecofantasy instead of ecomagination, promotes its jet as bird friendly.

A much more honest portrayal of the relationship between birds and jets is captured in a 1997 painting by Alexis Rockman titled *Airport* (figure 3.4). This one goes where Jack Griggs's image could not, given Griggs's responsibility to render birds identifiable in his field guide. Rockman, born in 1962, is a contemporary painter living in New York

City with a vast body of work providing frequently bleak and apocalyptic views of the effect of humans on animals, plants, and the environment. Replete with recognizable species of plants, insects, animals, and birds, much of Rockman's work is informed by the representational practices of zoology, entomology, natural history, and botanical illustration. As Rockman remarks in one interview, "I'm not a scientist, but I'm interested in generating credible scenarios that work on a number of levels, and I'm also interested in hijacking the language of credibility for my own purposes" (Rosenbaum 2001, 2). Although deploying aspects of scientific representation, Rockman's paintings frequently critique science and technology, questioning such things as the genetic modification of plants and animals and the role of technology in contributing to global warming (Rockman 2003).

In *Airport*, these themes come together as the carnage of bird strike is captured and dramatized in ways that make identification of some birds impossible. The airport in Rockman's painting is as abundant as any wildlife sanctuary, replete with several species of gull and an alert-looking rabbit. Mature and immature Laughing Gulls (with their characteristic black heads and wing tips) fly in the foreground, while a whirring jet engine dominates the center of the image. The engine hangs in space, having just ingested the bodies of one or more birds. In the midst of being sucked through the speeding blades of the engine, the bird bodies are being ripped apart. Blood splatters across the rotor blades. Although the bird or birds that have been consumed by the jet engine have been made indistinguishable, the ones that are still alive are distinctly recognizable as individual species.

In much the same way Griggs's image makes the claim that airports are an established feature in the lives of larks, Rockman uses specific, meticulously rendered images of birds to metonymically lend credibility to the claim that jet engines pose a danger to birds. Unlike GE, a company invested in making consumers ignore the relationship between jets and birds, Rockman publicizes what jets do when they come into contact with birds. In the style and subject matter of this painting, he challenges a tradition in modernist painting (represented by Charles Sheeler, for instance), in which modern technologies are aestheticized and represented as benign.

I have described Jack Griggs's image of Horned Larks at the Oakland

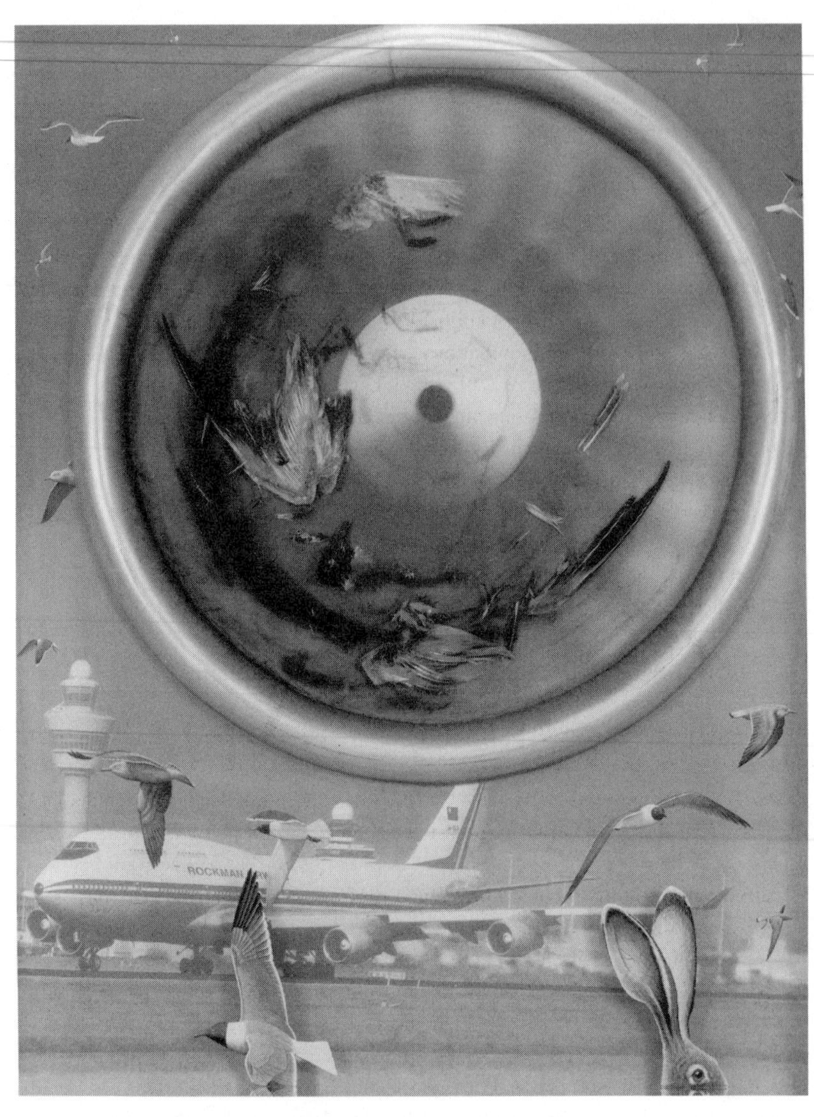

Figure 3.4. Alexis Rockman, *Airport*, 1997, Envirotex, digitized photograph, vacuum-formed Styrofoam with aluminum finish, oil paint, plasticene, Laughing Gull specimen on wood, 56 × 44 × 4½ inches. Collection Rochelle Lehman. Reprinted in *Alexis Rockman, with Essays by Stephen Jay Gould, Jonathan Cary, David Quammen*. New York: Monacelli Press, pp. 206–7. Courtesy Alexis Rockman.

International Airport as somewhat ambiguous, suggesting both the threats larks face and their resilience in the context of human technologies. Contrary to most other field guides, though, Griggs makes clear that Horned Larks and other species of North American birds do not live in pristine, unadulterated environments. For him, nature is a space of conflict and tension inhabited by humans and birds. GE's remix of Audubon's original (figure 3.2) attempts to erase any sense that jet aircraft impinge on birds and their habitats, and GE's message relies, in much the same way as Griggs's does, on the ostensibly fact-based nature of Audubon's original.

Rockman's *Airport* sustains multiple readings as well in that living birds and a rabbit thrive in the context of the jet aircraft while one or more birds have been sucked up by the jet engine, but the overall message of Rockman's painting is much more totalizing: jet engines not only kill birds, they make them indistinguishable. In Griggs's field guide and GE's ecomagination ad, species remain identifiable in the context of human technologies, suggesting that birdwatchers can make their way through a densely technologized world. In Rockman's painting, the ingested birds become unrecognizable upon encountering human technology, putting an end to any attempt at species identification. In this way, Rockman both creates the metonymic relationship between factually depicted birds and the destructive jet engine and disrupts it, showing how the technology is one that can erase not only the lives of birds, but the prospect that birdwatching may continue in a densely technologized present.

BIRDS AND POLLUTION

As briefly mentioned in chapter 2, it is not uncommon to hear gulls referred to as "trash birds" or "rats with wings," and such negative attitudes toward them can be professed by nonbirdwatchers and birdwatchers alike. Given this figurative association of gulls with trash, it is doubly appropriate that, in Griggs's field guide, Glaucous, Iceland, Thayer's, and Herring Gulls are pictured at an overflowing trash dump. I say doubly appropriate because gulls are not only called trash birds—many species of gulls live and feed at trash dumps.

In Griggs's representation of gulls, a Glaucous Gull in winter plumage consumes a piece of garbage, and around it, several other gulls perch on or fly over a teeming pile of waste. Newspapers, plastic six-pack holders, cardboard boxes, and plastic garbage bags are strewn about the scene. Even with all this trash, this image of gulls in Griggs's field guide remains every bit a classificatory field-guide image, with Glaucous, Iceland, Herring, and Thayer's Gulls separated from one another on two adjoining pages and presented in profile for the sake of comparison. To reveal a full range of field marks, the gulls are pictured standing and in flight, and particular attention is given to important field marks such as the color of wing tips, beaks, and legs. At the top of the page, disembodied heads of the four species highlight the unique combinations of distinguishing features found on gull heads alone (shape and/or color of beak, eye, and skull). Borrowing from a convention in natural history illustration in which multiple heads and other disembodied parts are often combined on a single page (Blum 1993, 321–22), Griggs's image asserts that to know gull heads is to know gulls.

Although the image aims to help readers differentiate similar-looking gulls, it also presents a jarring reminder of the role of human waste in the lives of birds. Griggs gives an ultrarealistic glimpse of one of the "natural settings" many gulls live in. Avid birdwatchers know that gulls can be found at landfills, and Griggs departs from other technical field guides by openly acknowledging this fact.

Like Griggs's representation of Horned Larks at the airport, the image of gulls at the dump can be read several ways. It calls into question the proliferation of human waste and how it is handled, displaying just how densely polluted some avian landscapes can be. And again like Griggs's image of the Horned Larks at the airport, the image represents gulls as resiliently able to carve out new niches in modified landscapes. Although the presence of garbage in the image may be jarring, there is no indication that the gulls are any worse off for their trash-dump surroundings. Even in a field guide, where the main goal is to provide exemplar images of every species, a somewhat sickly or deformed bird could be illustrative; it is not uncommon to see a gull with a withered or otherwise useless foot, for instance. But Grigg's gulls are thriving.

Because none of the birds appear diseased or unwell, the image implies, perhaps unintentionally, that trash dumps are adequate habitats for these birds. In terms of birdwatching, this image disrupts the notion, promoted by mainstream field guides, that birds live only in pristine habitats. Birdwatching is typically represented as a rarified practice and chance to "get away from it all," but Griggs's image says something else: for a birdwatcher to seek out gulls may mean braving a malodorous garbage dump.

Patricia Yaeger writes, in an essay about what she calls *rubbish ecology*, that "postmodern detritus has unexpectedly taken on the sublimity that was once associated with nature" (2008, 327). In this sense, trash is natural and even desirable. In another sense, the gulls are part of the trash. More traditional field guides promote and imagine birdwatching encounters as taking place in parks, mountain ranges, and around large bodies of water. Griggs's field guide cuts through that point of view, asserting that if a birdwatcher wants to find "all the birds of North America" (as Griggs's title suggests), he or she must be willing to hold his or her nose and brave, possibly even enjoy, certain toxic, polluted landscapes.

In another Rockman painting, *Concrete Jungle V* (1993), the notion that toxic pollution is harmless to birds is directly challenged. Even more important, Rockman questions the extent to which "looking" for toxicity and its outcomes amounts to a viable mode of examination. In Rockman's painting, which is part of a series, an active city looms in the background while a variety of animals and insects live out relatively healthy lives in a polluted trash lot. Fitting with Yaeger's notion of rubbish ecology, this painting can be read as an act of "saving and savoring debris" (2008, 329). Several teeming piles of garbage emit brown columns of smoke or gas while an array of biodiversity thrives around the garbage. Plants grow, there are insects and rodents, and two moderate-sized mammals fight or play (it is hard to tell which) near the garbage piles.

As Katherine Dunn remarked in *ArtForum*, "In [Rockman's] 'Concrete Jungle' series, 1991–94, he examines the secret gutter and garbage world of creatures adapted to life on the human periphery" (Dunn 1996, 73). Indeed, many animals seem to have adapted to this wasteland. Two distinct species of gulls circle in the air, for instance, one with dark plumage on the tops of its wings and the other with lighter wings but dark wing tips. Although identifying the species is difficult given their relatively small size in the image, Rockman takes care to portray the gulls as specific species with different plumage, leg color, and shape of outstretched wings. The image hinges, however, on a domestic cat in the foreground seen drinking from a pool of water that has been tainted by a tipped-over box of rat poison. The rat poison is, surprisingly, the only obviously toxic and clearly damaging element *that can be seen* in the filthy and polluted landscape.

Examining the image up close, one finds that the cat's skin is infested with parasites and its fur is falling out. Although birds soar in the half-sunlight, a possible marker of this landscape being healthy, close scrutiny of the cat reveals that toxicity and its ill effects are not always evident if one merely looks from afar. Only when printed on the side of a box of rat poison is toxicity made visible. Said another way, merely *watching* an environment, as birdwatchers do, can result in oversights in terms of the growing environmental crisis of toxic pollution.

Looking is not always seeing in Rockman's *Concrete Jungle V*, a painting that adopts a number of inversions: rat poison injures a cat, and the "concrete jungle," which is traditionally used to describe a city, exists far from the metropolis. Rockman's nature is a thoroughly altered and trashed one, but by no means is it a lifeless place. In fact, searching for birds in this maze of trash and crumbling concrete-block structures could have a certain appeal as insects, birds, and animals of all kinds have taken to this habitat. For Rockman, nature is what is now and what is real, not what is imagined as pure or untouched by humans.

Unlike mainstream field guides, which present birds in sterile surroundings or none at all, Rockman insists that seeing birds in general and "trash birds" in particular cannot happen without seeing the garbage that surrounds them. Even to the extent that the species of gulls in his painting are obviously distinct from one another but not entirely easy to identify (the gull with the dark wing tips could be a mature Herring Gull, for instance), Rockman portrays an alternate version of what a realistic encounter with birds might amount to.

In my reading of Griggs's image of gulls at the dump, I insisted that one way of interpreting the image is as a statement that human pollution can provide an adequate habitat for a range of birds. Although Rockman confronts this logic, questioning the visibility and viability of toxicity, another print ad from GE's ecomagination series endorses the idea that pollution is harmless to birds. In this ad, GE's illustrator has added two smokestacks to another Audubon original, "Glossy Ibis." Although Audubon acknowledged a human presence in the lives of birds by including a farm in the background of his painting—a motif in Audubon's work discussed in more detail below—the GE ad intensifies the effect of that presence with the addition of the smokestacks. Like GE's ad featuring "*Lessus Pollutantus*," this image promotes a corporate vision of GE as a clean-and-green corporation; it also supports the notion that looking is knowing when it comes to evaluating the effects of toxic environmental pollution.

Like the GE ad with the anachronistically inserted airplane, the smokestacks appear to be innocuous and almost "natural," looking as if they have sprung forth from the ground like trees. It is easy to overlook but worth noting that in GE's image of the Glossy Ibis the smokestacks exist, but the coal-burning power plant that would typically accompany them does not. Because of this absence, the smokestacks not only appear harmless, but are portrayed as divorced from their function, which is to belch steam, smoke, and sulfur dioxide into the atmosphere. As peaceful as the ibis in this bucolic, healthy landscape, the smokestacks assert a vision of a planet unchanged since the early nineteenth century when Audubon first painted these birds.

All the images discussed in this section—the gulls at the trash dumps in Griggs's *All the Birds* and Rockman's *Concrete Jungle V* and the GE-concocted Audubon—naturalize garbage and pollution in the lives of birds. For Griggs and Rockman, seeing birds in such settings is characterized as an important environmental lesson that is integral to understanding the scale of human impact on avian environments. Although GE's "Natura Harmonia" image of the Glossy Ibis also naturalizes the presence of the smokestacks, it attempts to assert that smokestacks contribute nothing but their presence—certainly not smoke—to the lives of birds. In their own and different ways, Griggs and Rockman attempt to bring about a kind of reconsideration of the trashed-out environments humans have made for birds, whereas the GE ad promotes a full-steam-ahead kind of attitude in which new technologies are portrayed as fueling industrial growth while conserving nature.

In terms of what picturing birds in trashed-out or otherwise polluted sites says about birdwatching, Rockman's rendering of toxicity as largely invisible certainly troubles any belief in bird*watching* as a sufficient mode of environmental assessment. All the images discussed in this section also disrupt the idea, propagated by mainstream field guides, that birdwatchers are the only ones who notice, identify, interact with, and discover birds. Field guides are predicated on a belief that special training, provided by the field guides themselves, is needed to know birds and that without this training birds might not be identified or even seen.

These images imply something quite different: birds travel through some of the most common and filthy areas on Earth. If many species of birds can be found in areas as unexotic as a trash dump, they have surely been seen by a range of people—not just birdwatchers—in everyday life. In this way, images of birds amid trash and pollution redefine what avian environments are and what kinds of people most often inhabit them. In the more mainstream field guides discussed in chapter 2, one gets the sense that birds are special and, as a result, live only in beautiful, special places, which makes birdwatchers special for seeking them out. In these images, one gets a different idea, one in which birdwatchers might be more akin to dumpster divers seeking out what has already been seen and passed over. The importance of the birder-as-dumpster-diver framework is that it never ceases to insist that the areas where birds live are less often pristine and more often toxic in some way.

BIRDS AND PEOPLE

At two points in Jack Griggs's field guide *All the Birds of North America* (1997), people are placed in the scene with birds: in one image, a birdwatcher can be seen looking at a group of plovers in a sandy estuary (56), and in another, a bundled-up group of nine birdwatchers is shown watching mergansers and grebes on a cold winter day (38). By placing images of birdwatchers in a birdwatching field guide, Griggs inverts the way field guides focus exclusively on birds. By placing these birdwatchers inside the frame of images that are ostensibly about birds, Griggs's *All the Birds* becomes self-consciously about birdwatching.

Although the authors of the first birdwatching field guides, published in the 1880s and '90s, used discussions of birds as an opportunity to comment on people and society, the technical field guides that gained prominence in the 1930s and are still popular today unselfconsciously maintain birds as the sole object of birdwatching. Griggs breaks from this tradition by commenting on birdwatching and its trappings by creating a scene populated by both birds and birdwatchers.

One way of reading the representation of this group of nine birdwatchers is as an implied critique of the effect that large groups of birdwatchers can have on birds and their environments. The problems these large groups can pose are not unknown in the birdwatching community. The American Birding Association has responded to complaints about groups of birders damaging habitats with its "Code of Birding Ethics," one portion of which stipulates that "group birding, whether organized or impromptu, requires special care" (American Birding Association 2009). Given the abundance of birds in the image and their apparent calm, however, Griggs's portrayal of the nine birdwatchers watching mergansers seems less a critique of mass birdwatching than a thoughtful inversion of the unrelenting focus on birds in field guides.

Here, Griggs characterizes mergansers and birdwatchers as both worth looking at and attending to. The group stands at the water's edge toting expensive gear: tripods, spotting scopes, and binoculars. From right to left, one birdwatcher points with an outreached index finger, another looks through a field guide, and the rest peer through binoculars and spotting scopes. They stand in a group, watching birds and ostensibly socializing in the way that birdwatchers do on National Audubon Society field trips, for instance. Three of the birdwatchers look directly out from the page at the reader, breaking the gaze that is otherwise trained exclusively on the birds. By looking out from the page, the pictured birdwatchers redirect the focus onto the reader and, by relation, themselves as birdwatchers who use field guides.

By associating birdwatching with high-tech birdwatching gear, the image also paints a picture of birdwatching as a commodified environmental practice. These people are not birdwatchers out looking at birds through simple binoculars, as many do, or with their naked eyes, but birdwatchers toting large scopes and tripods. In addition, notice how the image reduces birdwatchers to a set of their own identifiable field marks: the field guide, the scope, the binoculars, the location, and so on.

Granted, birdwatchers have long been parodied as oddballs who group up and trudge through natural areas in search of birds. In the *New Yorker*, for instance, cartoons have steadily appeared, since the 1940s, characterizing and caricaturing birdwatchers.[7] In addition, books such as Joey Slinger's *Down and Dirty Birding* (1996) and Simon Barnes's *How to Be a Bad Birdwatcher* (2005) poke fun at and celebrate birdwatchers. Although

Griggs's image has some element of this humorous critique of birdwatching, more important in the context of Griggs's field guide is how it inserts birdwatchers into the scene of birdwatching. In other words, the image of birdwatchers watching mergansers in Griggs's field guide introduces birdwatchers into the ecologies and taxonomies they study.[8]

In Audubon's original image of the Glossy Ibis, Audubon acknowledged a human presence in the lives of birds by including the small farm in the background. He included similar elements in several of his images. In his painting of the Golden Eagle, however, painted in 1833 for his collection of prints *Birds of America*, Audubon went one step further in including a small human figure in the background of the image. While the Golden Eagle soars through the air with a white rabbit impaled on its talons, a person shimmies across a fallen log with a dead bird strapped to his or her back.

Audubon's inclusion of a human figure in his painting of the Golden Eagle has drawn the attention of Audubon scholars such as Christoph Irmscher and Jennifer Baker in part because Audubon's printer, Robert Havell, removed the image of the person when the painting was transferred into an engraving and made into prints for the double-elephant folio edition of *Birds of America* (1840). Irmscher, in *The Poetics of Natural History* (1999), describes the human figure in Audubon's painting in this way: "In the left corner of the painting we see a fallen tree stretching from one rock to another, across which a man is slowly and laboriously making his way with the corpse of a bird (most probably another Golden Eagle) as well as a gun strapped to his back" (227). For Irmscher, the human figure is Audubon himself, making the painting in part about Audubon's practice of procuring birds for scientific study and illustration. Baker, discussing this same image, argues that "in the distance a woodsman—a version of Audubon himself—crosses a chasm by way of a fallen tree with a slain eagle and gun strapped to his back. In one sense the woodsman and flying eagle are analogous hunters. In another sense, however, the woodsman cannot be fully integrated into the natural world. The tiny figure, which was removed by Havell for the engraving, seems out of his element, groping his way across the log while the eagle effortlessly defies gravity" (Baker 2006, 4). Audubon's inclusion of the human figure is much like Griggs's inclusion of the nine birdwatchers in that both images openly acknowledge that birds live amid a human presence. But it is much more than that.

In Audubon's image, the work of procuring the specimen is revealed by the presence of the human figure, and in Griggs's image, the act of procuring an accurate identification is revealed by the birdwatchers in the scene. Both Audubon and Griggs have a host of other images with no people in them, but by including these human figures in their larger projects, they gesture to what it takes to procure a specimen, in the case of Audubon, and an identification, in the case of Griggs. The people are small in the images—small enough to be removed by Havell in the case of Audubon's work—but their presence is huge, adding a self-conscious frame to the images that aims to keep the focus not only on birds, but also on the people who seek them.

BINOCULAR VISION INVERTED

The images discussed in this chapter produce the distinct sense that by the time a birdwatcher sees a bird, that bird may have already narrowly avoided a collision with an airplane, dined on garbage, breathed in airborne pollutants, and been spotted by other birdwatchers out looking at birds with expensive equipment. From this perspective, an individual birdwatcher may be the last, not the first, to see a bird.

In previous chapters, I described binocular vision as a way of seeing and thinking about birds that has been constructed through relationships among birdwatchers, field guides, and birdwatching as an environmental practice. As Jeffrey Karnicky eloquently puts it, "Bird watching accommodates birds to a human visual apparatus, to allow bird watchers to differentiate birds by species (and subspecies and population), age, and sex" (Karnicky 2004, 254). This human apparatus, to use Karnicky's language, has been historically and culturally produced over time and has increasingly involved seeing and thinking about birds as detached from the physical, political, and ideological worlds that greatly affect them. Mainstream, technical field guides encourage birdwatchers to identify birds, but in doing so, those guides encourage a number of missed identifications and oversights.

What is important about the representations of birds discussed in this chapter is that they present a technically minded form of birdwatching as compatible with, not exclusive from, social and environmentalist commentary about birds and birdwatching. Technical field guides are largely based on the belief that learning to identify North American birds requires so much focus on the technical ins-and-outs of field marks and bird identification that there is no room left for commentary on such things as the perils of jet engines and pollution in the lives of birds, or even for reflection on birdwatching as a mass-culture, commodified niche market.

Mainstream technical field guides seem to assume that discussing such "social" issues has nothing to do with bird identification and even detracts from the scientific, objective nature of birdwatching. However, as the images discussed in this chapter show, a whole range of ideas, insights, commentaries, and points of view can be included as part of precise, taxonomic imagery about birds. In fact, such features make bird-watching more interesting, thorough, and environmentally engaged. As Alexis Rockman's *Concrete Jungle V* makes evident, the visual apparatus of birdwatching is in many cases a failed mode of examining environmental toxicity, but it can be one mode of inquiry, Griggs insists, that is at the very least attuned to the larger environments surrounding birds.

In chapter 4 I move to a discussion of how field guides have recently jumped from the technology of the book into a variety of portable electronic gadgets and interactive websites. This shift has brought about a range of changes in terms of how birdwatchers use field guides and what relationships with the environment networked field guides tend to foster.

CHAPTER FOUR

Technojumping into Electronic Field Guides

Thus far, I have made the case that field guides are based on a range of environmental assumptions with environmental implications. Field guides, though, are changing. Birdwatchers can now buy field-guide applications for personal digital assistants (PDAs) and the iPhone, and numerous online field guides provide detailed, interactive resources about birds. A portable gadget called the Song Sleuth can even identify birdsongs automatically by comparing songs that it "hears" to those in its onboard database. Field guides are becoming automated.

Although not yet available for birdwatchers, a handheld electronic telescope called the SkyScout automatically identifies stars and constellations using GPS (www.celestron.com/skyscout), and a team of researchers at Columbia University have developed a pair of virtual reality goggles that allow users to identify tree leaves simply by looking at them (White, Feiner, and Kopylec 2006, 2008). Fully automated bird identification devices of this kind cannot be far off.

In this chapter, I take account of this move, noting that the genre of the field guide has recently jumped technologies, moving into a range of new, electronic platforms. These new wired field guides reveal a lot about how practices of bird identification are changing to become both more commodified and diversely networked. More important, though, is that comparing electronic field guides to their print counterparts demonstrates just how much print guides, because of the affordances of book technology, have long promoted bird*watching* as a set of visual practices.

Field identification and environmental study have developed, in the pages of modern, technical, print field guides, as visual practices that, as a result, only partially render and imagine their frequently noisy and always moving objects of study. Wired guides reveal, complement, and in some ways challenge this partial and largely visual approach to representing birds. The new wired guides render bird identification in a way that is as much about listening to birds as looking at them; wired guides also position information about birds in new networks that incorporate participant-generated content. Although many things are unchanged about birdwatching with a wired guide, the recent technojump amounts to another step into the gadget and commodity culture for birdwatchers who go digital.

In addition to discussing how ways of seeing are in part dependent on the technologies birdwatchers use, this chapter connects with existing scholarship on how genres such as field guides change over time. Randall Popken, for instance, has shown how the everyday genre of the résumé evolved dramatically from a format much like a letter to the list of terse descriptions we know today (1999). Thomas Miller has shown how the genre of the essay evolved and changed over multiple centuries, still bearing traces of its roots as an "essay of taste and manners" (1997). In some of the more influential scholarly collections on how genres work and change (Freedman and Medway 1994; Coe, Lingard, and Teslenko 2002), genres have been studied as if they only exist in single technologies. Field guides, though, bridge multiple technologies.

Playable birdsongs are one of the main new features in these wired field guides, and such recordings are being incorporated without restructuring the overall emphasis on bird identification found in the guides. A "history of the book" regarding field guides can and must go digital, as the image of a birdwatcher, standing in a meadow and paging through a print field guide, is becoming increasingly antiquated. Robust guides are now on smart phones, online, *and* in print.

Exploring the technojump to portable electronic and online field guides yields an understanding of these new-media guides and thus provides insight into the role of new-media gadgets and texts in environmental encounters, but it also reveals the long-standing significance that book technology has had in textually sponsoring bird*watching* (as opposed to something that might be called bird*listening*). The technology of the book, which is more amendable to representations of the visual than the audible, has had a significant role in shaping largely visual representations of birds in field guides. Since roughly the 1930s, print field guides have developed into increasingly visual texts. With this move to the visual in print field guides has come a simultaneous move away from the audible. Now electronic guides do both.

This chapter begins with a short history of efforts to visualize bird vocalizations[1] on the printed page. It then moves to a discussion of portable electronic gadgets, applications for PDAs, and two popular online birdwatching field guides.

MAKING FIELD GUIDES SING: A BRIEF HISTORY

Since the publication of the first birdwatching field guides in the 1880s and '90s, a variety of methods have been used to describe, reproduce, and visualize bird vocalizations in printed field guides. Vinyl records of bird vocalizations were available since at least the 1930s (Brand 1934), but they could be used only within the home and included very little additional information about the songsters they featured.

The most comprehensive attempt to capture bird vocalizations on the page was by F. Schuyler Mathews in his remarkable 1904 field guide, *Field Book of Wild Birds and Their Music*. In this guide, Mathews dispenses with how birds look, turning instead to how they sound. He puts the songs of North American birds on musical scales, highlighting each bird's musical qualities and including lyrics for each song. Birds in this system of representation are imagined as part of what could be called "nature's orchestra" and in that sense are akin to feathered musicians. When it comes to representing birdsongs on the printed page, Mathews upholds the fidelity of the musical scale by critiquing the method of translating birdsongs into writing.

Comparing how the Red-winged Blackbird's song was translated into writing by different nineteenth-century authors, Mathews notes: "Various writers interpret the syllables differently. Emerson's opinion is that 'The Redwing flutes his "O ka lee."' Mr. Chapman makes it 'Kong-quer-ree'; William Hamilton Gibson, 'Gl-oogle-eee'; and yet another writer, 'Gug-lug-gee'" (Mathews 1904, 54). After sorting through these variations, Mathews decides on the highly glottal "Gug-lug-gee-e-e-e-e-e-e-e-e!" as a way of describing the species' song. Ambitious as it is, Mathews's musical approach to capturing birdsongs never really caught on, and subsequent field-guide authors reverted to using phonetic transcription to represent bird vocalizations.

As David Rothenberg points out in his detailed study of birdsongs, in "the 1940s researchers at Bell Telephone Laboratories in New Jersey invented the sound spectograph, otherwise known as the sonograph, at first for the purpose of identifying possible criminals by their voiceprints" (Rothenberg 2006, 61). Not surprisingly, ornithologists immediately began using sonography to study bird vocalizations. Rothenberg stresses the significance of sonographs in data-driven bird-song research by stating, "At last, the strict details of the song could be objectively rendered on the page, and we could approach an exactness far beyond mnemonic musing and quirky squiggles" (63). By the mid 1960s, sonograph representations of birdsongs made their way to birdwatchers via their incorporation in Chandler Robbins, Bertel Bruun, and Herbert Zim's field guide, *Birds of North America: A Guide to Field Identification* (1966).

Much as special knowledge was required to read the musical scores in Mathews's field guide to birdsongs, though, it took a special literacy to make sense of the sonographs in this field guide. As a result, neither method of representation was intelligible to all users. Sonographs lent Robbins, Bruun, and Zim's field guide an aura of scientific precision, but this mode of visualization did not make it easy for readers to "hear" what birds sound like. Like Mathews's musical scales, the sonographs never really caught on, and subsequent field-guide authors reverted once again to phonetic transcriptions of birdsongs. The Red-winged Blackbird's song is described as *Conk-ca-ree* in Jack Griggs's *All the Birds of North America* (1997) and *kon-ka-reeeee* and *opPREEEEEom*, depending on the population, in David Allen Sibley's *The Sibley Guide to Birds* (2000).

Recently, some hybrid forms of new media have begun incorporating playable recordings of bird vocalizations. The simplest of these is represented by Les Beletsky and Jon L. Dunn's book *Bird Songs: 250 North American Birds in Song* (2006), which is a narrative field guide equipped with a plastic push-button playback device integrated directly into the back cover. The reader/user of this book can flip through the pages, looking at images of various species and reading facts about them, and then push a button to hear one or more recordings played through the book's tiny speaker. Although this book is more of a novelty item than a usable field guide, Beletsky and Dunn's *Bird Songs* exemplifies the recent move to transcend the confines of the printed page to more vividly present birdsongs.

PORTABLE ELECTRONIC FIELD GUIDES

Of the many new electronic devices on the birdwatching scene, two of the simpler ones focus entirely on birdsongs. IdentiFlyer from the company For the Birds and Coleman's Birdfinder (manufactured by Polyconcepts USA, Inc.) are rudimentary in the sense that they do not have electronic displays like other birdwatching gadgets, but they are important in being able to play recorded birdsongs at the touch of a button. Both gadgets are small, intuitive, and forest green (suggesting outdoorsy-ness), and the lanyard on Birdfinder indicates that it is meant to be taken out into the field.

As one of the testimonials on IdentiFlyer's website suggests, however, consumers may be using IdentiFlyer and gadgets like it more as instructional tools or novelty items than to directly mediate their encounters with birds: "I was thrilled to see and purchase your Birdsong IdentiFlyer quite some time ago at Wild Birds Unlimited. Knowing it would be a great item, I bought the only other two cards available at the time. I have thoroughly enjoyed it myself, with my husband, and with my grand kids. This is a great way to really show them [in a moment] what a particular bird sounds like" (Birdsong IdentiFlyer 2007). Although it is unclear that many birdwatchers are actually using IdentiFlyer and Birdfinder to help them identify birds, both portable electronic devices exemplify how new kinds of gadget-guides are emerging that integrate playable bird sounds while maintaining the long-standing emphasis on accurate bird identification.

To identify the Yellow-breasted Chat or Blue Jay using IdentiFlyer or Birdfinder, one inserts the appropriate card or disc (IdentiFlyer has fourteen song cards, each sold separately) into the gadget, pushes the button beside the image of the species in question, and listens to the song. To use the recorded song to attract the species in the field, a practice used by some birdwatchers, the button can be pressed as many times as necessary. The user of a print guide might read a phonetic transcription of a bird's song and try to sound or whistle it out. The user of a portable electronic guide listens to a repeatable sound file in an attempt to match up what is heard on the device with what an unidentified bird might sound like.

These different processes involve the consultation of reference materials at the center of the exchange, but portable electronic guides do more than present birdsongs in new ways. These new gadgets present users with mechanical, battery-operated, on-demand singers that stand in for birds while being able to attract them. In much the same way that print field guides rely on a single image of a species to represent all members of that species, no matter how much individual members of a species vary, portable electronic devices presume that all members of a species are knowable through a single sound recording. This set of transformations and reductions adds an air of precision to birdwatching with an electronic guide, where studying birds with such a device is streamlined by the definitive realism of the reference technology.

Software applications for Palm OS and iPhone take the rudimentary capabilities of For the Birds's IdentiFlyer and Coleman's Birdfinder one step further, integrating recorded sounds with stored digital images and written species descriptions, all searchable within a single handheld application. There are many of these new birdwatching applications, including ones with such titles as Winged Explorer for BlackBerry, PocketBird 1.0 for Palm OS, iBird Explorer for iPhone, Chirp for iPod, and National Geographic's WhatBird for PDA. eNature.com sells a version of this software, its Handheld Guide to Birds (for the Palm OS), that amounts to a visually and audibly rich, interactive electronic field guide. Handheld Guide to Birds comes in Premier and Lite versions, with the Premier version shipped in fifty different configurations, one for each U.S. state. If you live in Oregon, for instance, your Premier software will come loaded with information on that state's 345 native birds. Washington residents get fifteen species fewer for the same price. (Lite users pay $3.99 for information on twenty to one hundred species.) And as eNature.com is quick to point out, profits from the sale of the software go to eNature.com's related nonprofit, the National Wildlife Federation. In providing revenue for the National Wildlife Federation, the guide is akin to other popular print guides sponsored by large environmental nonprofit groups. eNature.com's software runs on a portable electronic device, enabling a Palm Pilot to become a searchable, interactive field guide with touch-screen navigation. Users can view range maps, digital images, listen to birdsongs, and read about birds using this software.

This electronic field-guide technology replaces the physical act of flipping pages (in a print guide) by introducing new search protocols. Instead of paging through a guide, a user of this guide would conduct a Boolean search for multiple variables linked by "and" inclusion terms. Photographs replace paintings in this guide, as realism and a sense of precision are integral to electronic guides. There is little "new" information about birds that one could not find in a print guide, aside from the incorporation of recorded songs; the content of electronic guides is largely repurposed from print sources and represented in the e-guide. With this field-guide architecture, by clicking on the "map" tab, then back to "description," and next over to "images" the user can move between informational screens in a mode similar to searching for information online.

In terms of how these guides function as cultural devices or what they do socially, all the portable handheld devices I have discussed work in much the same way as print field guides. In this sense, they are wired versions of the technical guides discussed in chapter 2, focusing on field marks and the visual representations of birds. Like their print counterparts, these guides sort birds into categories and highlight species differences for the purpose of identification. At the same time, significant differences exist: these portable electronic devices are much smaller than most print guides, making them even easier to carry, and they are searchable in different ways, making the guides physically navigable not by turning pages but through clicking or tapping on hypertext links. Information is presented on a small screen, of course, and the integration of sound files offers users the ability to hear birds instead of reading about their songs.

Although simple gadgets such as IdentiFlyer and Birdfinder may always be novelty items for birdwatchers, field guides for Palm OS and iPhone are poised to make viable inroads into the daily reference work of birdwatching. Using a field-guide application for iPhone such as the one simply titled "Birds" is easy and intuitive, allowing the user to do things such as save sightings and run through visual and audio quizzes. Whereas print guides continue to be limited in terms of how they connect users to information and what method they use to represent birdsongs, portable electronic guides offer the potential to overcome such challenges. Of course, following a Boolean search is not always desirable, and flipping through a print guide offers not only the potential for finding what one is looking for but the chance to stumble on something surprising.

Navigating an electronic field guide is much different from paging through a print guide in that e-guide navigation entails following a predescribed path toward a single target. Once variables are ruled out in a search, for instance, possible "suspects" are set forth by the software. Furthermore, the rhetoric of identification with birds, not only of birds, finds even less of a place in portable electronic guides than it does in their print counterparts. Portable electronic guides conflate new forms of high-tech (seemingly scientific) materiality with now-established traditions of scientific discourse. The high-tech materiality of these portable electronic devices accentuates what seems like a scientific, objective approach to representation and a way of relating with "nature" using expensive, battery-operated tools.

In this way, IdentiFlyer, Birdfinder, software for Palm OS, and iPhone applications exemplify how birdwatchers are under increasing amounts of commodity pressure constructing them as in need of high-tech gadgets. Most birdwatchers have long extended their bodily capacities of sight through the use of optics, and print field guides are another tool birdwatchers have used, but using these plastic, electronic devices—some equipped with LCD displays and all able to play recorded bird vocalizations—even further integrates mass-produced mechanized technologies into the fabric of birdwatching. Birdwatchers using these tools have consented to integrate expensive gadgets into their environmental practices, but as the title of Florence Merriam's *Birds through an Opera Glass* (1889) reminds us, birdwatching has always involved the coupling of the birdwatcher with gear. The tool of the opera glass replaced the tool of the gun for studying birds in the nineteenth century, and now the tool of the electronic gadget is posed to replace or at the very least supplement the tool of the printed field guide in the twenty-first century.

This idea of a heavily gear-laden birdwatcher has received some criticism from inside and outside the birdwatching community. Felton Gibbons and Deborah Strom write, for instance:

> Books, magazines, cassettes, video tapes, binoculars, scopes, feeders, bird fountains and travel all cost money, and the serving of birders has become a multimillion dollar business. As with any hobby, the apparatus mounts up, becoming a collection in itself. If the bird watcher has also taken up bird photography, his opportunities for amassing cameras, lenses, and filters are unlimited. One hundred years ago Celia Thaxter [nineteenth-century poet and bird lover] watched birds with her naked eyes and identified them by writing description of them to Bradford Torrey. Today the birder is overwhelmed with apparatus which threatens to take over the sport. (Gibbons and Strom 1988, 308–9)

In truth, Celia Thaxter's nature study was not entirely unencumbered—at the very least, the apparatus of scientific classification mediated Thaxter's birdwatching experience—but the vast range of consumer goods marketed to contemporary birdwatcher-consumers as "necessary" has certainly increased. The connection of human to manufactured device (be it a simple opera glass or expensive pair of DigiBinos) has always been foundational to the pastime, and new forms and scales of commodity pressure reify and extend the commodified and wired birdwatcher.

One way to think about portable electronic field guides is as being imbued with what Jean Baudrillard has called the sham object and the simulacral. For Baudrillard, commodity culture is a system of signs and language, and using various products means taking part in a kind of conversation that makes meaning with those products. What gadgets and sham objects[2] represent, for Baudrillard, is even more important than what they do in any practical sense (1994). An item of clothing is first and foremost a statement, in this way of thinking, and only secondarily something to cover one's body or use to keep warm. As sham objects that traffic in simulation and simulacra—recorded birdsongs played on small speakers—electronic handheld guides represent a continuation and intensification of the belief that encounters with birds and nature require commodities that will extend the body and produce authoritative, sanctioned identifications and recordable experiences. These devices construct effective bird identification as unlikely without the assistance of high-tech gear. Indeed, a thoroughly geared experience is constructed as necessary if one is to go out in search of birds.

But, still following Baudrillard, high-tech birding gadgets do even more. In his chapter "The Precession of Simulacra," Baudrillard uses the example of a Disney theme park to suggest that such a cultural phenomenon is not a mere production of a city, but indirectly the production of the notion that cities are able to be simulated (1994). A similar logic can be applied to what birdwatching's portable electronic devices semiotically produce: they make birds appear simulated and electronically replicable (even replaceable).

By integrating high-resolution digital images and quality sound files, applications for Palm OS and iPhone extend an ongoing insistence on the part of classificatory birdwatching technology that birds can be reduced to a taxonomic essence that can be captured and displayed through a reference device. That is the aspect of print field guides that Michael Lynch and John Law (1999) find so frustrating: namely, that even with species that are impossible in most instances to identify by sight (hybrids and subspecies are particularly challenging, as are similar-looking species), field guides continue to insist on the viability of Petersonian field marks. Portable electronic devices enact and imagine a desperate, needy kind of birdwatcher who cannot function without new technologies to help identify birds. Be that new technological device an unplugged, highly designed waterproof field guide or a portable electronic device, what is manufactured in the exchange, use, and circulation of these tools is both the image of a birdwatcher needing to mediate encounters with birds using classificatory technology and a hobby focused more and more dramatically on the simulation of birds in ways that simplify them for ready identification.

ONLINE FIELD GUIDES

Various communication technologies have long been used by birdwatchers to share information about rare bird sightings. Telephone calling trees and recorded rare bird alerts are still in place in some areas, helping birdwatchers network by telephone about the locations of rare birds in a particular area. The use of the telephone has since given way to messages about rare birds being sent via e-mail, Listserv, text message, and Twitter. The Internet now offers an even broader array of opportunities for birdwatchers to communicate with one another, including an applet called the eBird Rare Bird Google Gadget that notifies birdwatchers about recent sightings and includes a link to a Google map plotting the exact location where each rare bird was seen. Birdwatching tools of other kinds abound online. Birdwatchers who keep lists, for instance, can download the application Avisys to store and sort detailed lists of sightings. There are also several online field guides.

Both eNature.com and Cornell University's Lab of Ornithology host such online guides, and comparing them shows how both work toward the goal of making birds identifiable through slightly different kinds of web tools. A number of scholars, including Marshall McLuhan and Eric McLuhan, Jay David Bolter and Richard Grusin, and Lev Manovich, have discussed the emergence of new-media forms of this kind. Summarizing this scholarship, Kevin Brooks, Cindy Nichols, and Sybil Priebe write that "one of the fundamental principles of new media . . . is the principle that old media and familiar genres end up as the content of new media" (Brooks, Nichols, and Priebe 2004, para. 6). In a sense, that is what is happening with online field guides in that they look, feel, and work like print guides, accomplishing the objectives of print texts in new ways.

By making the content found in print guides the content for online guides, online field guides bridge a possible gap between a known technology and a new one via consistent forms of content. Said another way, when birdwatchers encounter online field guides, they know what they are looking at because they have already seen a print guide. All a user of an online guide needs to do is to figure out how to use familiar content in a new format.

The first thing to note about these new online resources is that both the guide at eNature.com and the one hosted by the Cornell Lab of Ornithology are parts of much larger websites. eNature.com is related to the nonprofit National Wildlife Federation and as such the two sites are somewhat bundled together. Similarly, the Cornell Lab of Ornithology's guide connects via hyperlinks to many other resources offered by the lab. They are not stand-alone guides, then, in the way that print field guides are, but rather networked online entities that connect via hyperlinks to an array of other materials and sites.

For instance, from the main eNature.com page, users can link to online field guides about "amphibians, birds, butterflies, fishes, insects, mammals, native plants, reptiles, seashells, seashore creatures, spiders, trees, and wildflowers" (eNature.com, n.d.). This fact alone—the linking of a bird guide with other kinds of field guides—represents how much more thoroughly networked online field guides can be in comparison to print guides.

I have argued that print guides about birds separate birds from their surrounding landscapes and pressing social debates. eNature.com's guide locates reference materials about birds amid a variety of other types of nature study (represented by guides about other natural categories), creating a more complex and imbricated view of the environment. Similarly, the Cornell Lab of Ornithology's online field guide connects to a thorough "All About Birds" page, locating bird identification amid many other aspects of bird study and ornithology. Such embeddedness, in the cases of Cornell's online guide and the lab's many other online resources, expands on the singularity of what was once exclusively a "bird guide." As part of larger sites and hypertext networks, these online field guides extend beyond the confining scope of print.

Searching for a species of bird using eNature.com's online field guide involves a three-step sequence. Like most hypertexts, however, this sequence can be circumvented to some extent, stopped by the user, or dropped in on midway via a link from elsewhere online. Taking the potential for variable forms of use into consideration, the three steps users of the site are encouraged to take are (1) select a type of bird from one of the fourteen types listed on an "all birds" link, (2) select an individual species from a sequence of roughly fifteen thumbnail images of birds with common and Latin names below each one, and (3) click to and view the species entry for that species. The species entry generally (although not in all instances) includes a resizable image, sound file, descriptive text, and links. At this point, the process can be repeated for another species (returning to step one in the sequence) or users can click to the next or previous species organized in a hypertext chain from one to the next.

Through this architecture, users are initially funneled from a field of many species to several to one, and upon reaching an individual species description users can either choose to be refunneled back through the same sequence or click to the next species. One cannot flip ahead five species at a time, as is possible in a print guide, but users can click around in the guide from entry to entry.

This branch-tree (as opposed to Boolean) search structure exemplifies how the repositioning of information from what exists in print guides makes for a somewhat different type of text online, particularly in terms of navigation. Whereas technical print guides are ordered compendiums of species descriptions (text and image) users flip through, eNature.com's three-step structure keeps the full volume of information contained in the guide below the surface of what the user can access at any one time. Searching a print guide is made possible via an index and various other access markers, and the architecture of print field guides facilitates casual armchair reading.

The species descriptions in the online guide at eNature.com can, for instance, be clicked through (users can choose to view "all birds" from the main screen), but that clicking is necessarily methodical and preconfigured by the guide, so such clicking around does not become a useful way to navigate the online text. Instead, the 1–2–3 search sequence is the predominant way the guide is organized, and in this way eNature.com presents itself as a database users reduce in size (by clicking "chicken-like marsh birds" or "swallow-like birds" and then selecting the next criteria) before seeing descriptions of species in that class. eNature's reader is positioned as a searcher in need of an instant, not just quick, identification solution.

Compared with print field guides, the species descriptions at eNature.com include more text and take up more space on the screen than their print counterparts—a luxury of the display space of the computer screen. Images and sound files are part of nearly every entry at eNature.com, as are banner ads (discussed below). In terms of design, the species descriptions are not unlike print guides: text and image appear alongside each other following a recognizable template, and the written descriptions all move from common name, to Latin name, to description, to discussion of "endangered status" (it is the National Wildlife Federation, after all), followed by habitat, nesting information, range, voice, similar species, and a short "discussion" about each species.

As this online guide shows, the physical constraints print guides pose are largely irrelevant online, making for more content and longer descriptions. Users of a book might be able to flip from one description to another more freely and easily than e-guide users, but once the species description is pulled up on eNature.com, it offers more information, inviting a user to stay longer and learn more. In this way, this e-guide is somewhat less of a quick-reference guide and more of an encyclopedic manual reminiscent of tomes about birds that were popular in the late nineteenth century (Coues 1872). That makes perfect sense, as eNature.com's online guide is not meant to facilitate bird identification in the field.

In that the eNature.com guide includes several types of advertisements, recontextualizing the genre of the field guide from print to the web has meant bombarding users with ad content in new ways. Make no mistake: eNature.com is a monetized website designed to generate revenue for the National Wildlife Federation. Most obviously, users are encouraged to "DONATE NOW TO THE NATIONAL WILDLIFE FEDERATION" via a link at the bottom of each species description. Embedded just beneath each species name and near the top of the page is also a revenue-generating link to the online retailer Amazon.com. Indeed, each page of eNature.com's online field guide is surrounded by a changing array of banner and fixed ads, all linked. If you need a hotel room, you can link at the right of the page to Orbitz.com. Are "Pest Birds a Problem?" You can simply call the "Complete Bird Control Product Line . . . 800-503-5444 for Expert Help! www.birdbarrier.com" (Black-footed Albatross 2007).[3] The content of the banner ads is automatically generated by the word *bird* in the web address of the page.

Print guides have always been revenue-generating devices—most recently for the nonprofit organizations that sponsor them—but because online guides have no subscription or purchase price, eNature.com demonstrates how important advertising dollars are to generating revenue, complicating and multiplying the messages present in the content of the guide. As a text encompassing an entire screen, this e-guide is centered on species identification and manifests this information by giving all species relatively equal amounts of content. But a banner ad such as the one for Birdbarrier.com becomes part of this text, and in this case the banner ad introduces a reminder (discussed at length in chapter 2) that all birds are not valued equally. Banner ads multiply the messages in this e-guide, splintering its central message to generate income in the economic environment of the web.

The online field guide at eNature.com shows how recontextualizing a genre from the technology of the book to another technological platform invariably leads to some alterations to the original content. eNature.com's recontextualized guide still privileges the work of bird identification while including new kinds of information (embedded sound files and resizable images) and a whole new dimension of commerce and commodification that is bundled along with the classificatory apparatus of the text. This guide no longer imagines the birdwatcher as an isolated hobbyist, but as an integrated hobbyist-consumer, complicit in using a guide replete with banner ads, but also potentially interested in birdwatching as part of a much larger set of nature-study pastimes.

The online field guide hosted by the Cornell Lab of Ornithology, called the Online Bird Guide, uses a different system of information management. The differences between Cornell's guide and the one at eNature.com exemplify how much online guides currently vary in the uncodified new arena of online field guides. If there can be said to be a naviga-tional sequence to Cornell's Online Bird Guide (and again, actual use of any website is sure to vary), it involves (1) either searching for a species name or browsing species by either taxonomy or "name and shape," followed by (2) viewing a species' page. At this point, a user can opt to (3) click on one of the informational options labeled identification, life history, sound, and video. Users can jump quickly to pages for other birds via either a pull-down menu or links to similar or related species. By offering a wealth of information, Cornell's Online Bird Guide is a resource that caters to experienced birdwatchers.

Similar to eNature.com's guide, Cornell's online guide includes such interactive features as resizable images and playable sounds; this integration of text, image, and sound is perhaps the defining feature of online guides.[4] Cornell's online field guide also adds a range map with plotted data of actual sightings, an element taken from print guides and enhanced. Interactivity is a key feature in this and other online field guides, allowing users to customize the experience of consulting Cornell's Online Bird Guide. Whereas print guides are geared for birdwatchers of different abilities (David Allen Sibley's guides are quite advanced; the Stokes's guides [Stokes and Stokes 1996] are for beginners), this site caters to both sets of users via tabs that offer different amounts of information about each bird. With this online guide, users can navigate the content so that it works for them. No print guide is this flexible or shaped in part by the user.

This interactivity demonstrates how online guides, like most appealing websites, give users things to do when they navigate to the site. Birdwatchers can click on and resize images, activate pop-up windows with sound files, play and replay those sounds, click over to a more "Detailed Page," capture images and text for downloading and future use, and choose from a range of links to birding-related topics. The guide is a tool that responds in the presence of the user.

The Cornell Lab of Ornithology's Online Bird Guide also has carefully linked citations for its images, sound files, and textual content, subtly foregrounding the authors of the guide's content. Although not all online field guides attend to citations as carefully, the incorporation of such information sets Cornell's online field guide apart from multiply-authored print guides such as Allen and Hottenstein's successful National Geographic Society *Field Guide to the Birds of North America* (1983), for instance. The images in Cornell's Online Bird Guide are embossed with each photographer's name, the recordist of sounds is clearly cited in pop-up windows, and at the bottom of each species description is a section titled "Sources used to construct this page" with source texts named and in many cases with links. Careful citation of this kind is not the norm at eNature.com, but is adhered to on the more academic Cornell site and other online field guides such as the popular and widely used Patuxent Bird Identification Info Center (Gough, Sauer, and Iliff 1998).

Although many print field guides stand behind the figure of a single author—Roger Tory Peterson, David Allen Sibley, and Kenn Kaufman—online guides provide multiply-authored content. Many print guides are created by teams of writers, artists, and designers, too, but what is different about Cornell's multiply-authored e-guide is that such creators are cited and linked to in the guide, recognizing this decentered authorship and emphasizing collaborative creation by a community of experts.

Why does the authorship of a field guide matter? Field guides created by a single author continue a tradition from early naturalists such as Alexander Wilson and John James Audubon, who conducted their own studies, did their own illustrations, and wrote in detail about birds. The recontextualization of the genre of print field guide to the Internet, however, changes the way singly-authored texts are privileged and seen as more authoritative.

By foregrounding and naming the multiple authors and sources that go into the creation of a single field guide instead of masking them, decentralized authorship takes on a new kind of authority. The U.S. Geological Survey's Patuxent Bird Identification Info Center (Gough, Sauer, and Iliff 1998) is a good example. It is an easy-to-use bird identification site that, like Cornell's online field guide, carefully cites the print field guides from which each entry borrows as well as the sources for images and sounds. This guide and others like it make clear that the content users are examining was not all originally composed for the web, nor is the guide produced by an artist/scientist in the mold of Roger Tory Peterson or David Allen Sibley. Instead, online guides recontextualize clusters of information from vetted and viable sources.

Multiply-authored texts of this kind present themselves as a sort of carefully curated "best of the best," and in constituting themselves in this way, they become akin to much of the other content users find online. It is not that authority is not granted to an e-guide by foregrounding multiple authorship, because it is; it is that authorship is expressly made multiple, and with online field guides, multiple authorship is the norm. Such authorial multiplicity mirrors the ways users of online field guides use online materials much more fluidly, moving from one guide to the next in search of information that is not necessarily associated with a specific cult of personality manifest in a Peterson, Kaufman, or Sibley.

The guides at eNature.com and the Cornell Lab of Ornithology's website represent somewhat different ways online guides have been used to accomplish what print field guides have long done: render birds identifiable. Online field guides are like their print counterparts in that they still focus on bird identification, still feature taxonomic images and text, and are still designed as quick-reference texts. With that said, there are several differences in the online guides. Remarkably, online field guides are resources one can listen to. Instead of reading phonetic transcriptions of birdsongs in online guides, just click on a button and let the recording or video play. Unlike portable electronic devices, and for that matter print field guides, online guides are not typically held in a birdwatcher's hand and taken out into the field to meet birds, but online guides do bring the birds to the user in multimodal, shockingly lifelike ways. Online guides are vast information centers, reference hubs, and new structures for presenting old forms of taxonomic thinking.

Binocular vision as I have described it is a visual system, and it includes certain oversights. As the birdwatcher focuses in on a bird and works to determine the species classification it is a member of, important features in that bird's surroundings may fall into the background and be considered less important than the bird itself. In this chapter, I have delved further into this issue of oversight in two ways. The first way was by making the point that it is in part because of the technological limitations (and virtues) of the printed book that watching birds (instead of listening to them) has been promoted as the best way for bird*watchers* to encounter and identify birds.

I would not go as far as to suggest that were it not for highly visual field guides people would not watch birds, nor is it the case that print field guides are the only reason for this turn to the visual in bird identification. It is the case, however, that the technology of the book has never been particularly robust at representing birdsongs.

This leads to my second point, which is about how new kinds of electronic field guides emphasize the important role of listening in bird identification. Although this newly emphasized attention to bird vocalizations may seem to disrupt the visual culture of birdwatching, the increased attention to sound also maintains the long-standing emphasis on species identification that is in part responsible for the other oversights mentioned in this book. It is far from the case, then, that electronic field guides are changing the face of birdwatching as long as we understand "the face of birdwatching" to be the core beliefs and practices of quick identification.

A classification-minded kind of birdwatching is ultimately enhanced and encouraged in these new guides, as a decontextualized identification-minded approach continues to be promoted by most American field guides to birds. Like print field guides to birds, portable electronic and online guides present distinguishing data in seemingly objective ways to bring about comparisons that result in naming and classification. Classes and classification maintain their primacy in these technologies; the core features of the genre remain intact. It is true, though, that these new electronic field guides are very different types of tools for identifying birds. New forms of interactivity between birdwatcher and field guide are now possible and audio files that could not be housed in books are now available in electronic field guides, but new field guides continue to be proponents of binocular vision.

At the beginning of this chapter, I mentioned a team of researchers at Columbia University who have created a "mobile augmented reality system" for identifying plants by their leaves. This technology demonstrates the extent to which electronic devices are introducing major changes to the ways we identify aspects in our environments. Increasingly, people use GPS-equipped maps for navigation, and new generations of digital cameras can automatically encode geolocational data onto every image. For birdwatchers, it means that an image of a bird can contribute to databases of where birds are. The identification of bird sounds has already been automated with the invention of the Song Sleuth, and although this device is a bit clunky in its design, more appealing song identifiers are sure to be released.

It is not hard to imagine a day in the not-so-distant future when a birdwatcher will be able to spot an unknown bird through a device that looks very much like a pair of binoculars and interact with an interface that fully or partially automates the identification of that bird by both sight and sound and then maps the bird's and the birdwatcher's locations before uploading a data file to the web. By downloading other bird data files, such a device could provide users with the likelihood of an identification, so a bird that looked like a Crested Caracara in New Jersey would be unlikely, but a Turkey Vulture would be likely. By making data more readily searchable and accessible to users in the field, the term *birdwatching* starts seeming antiquated.

This is not to say that bird books are not and will not still be important. Many birdwatchers love their guides, and many, myself included, collect them. In fact, the technology of the book has certain affordances that are lost in e-guides, such as flipping through pages or accessing a random entry. But e-birding is not in the future; it has already arrived.

It is only a matter of time before the gadgets birdwatchers use get more effective at not only identifying birds, but interfacing with vast databases and social networks of birdwatchers. In 2008, the U.S. Geological Survey published a vast archive of bird-sighting information to the web via the website www.pwrc.usgs.gov/bpp. The North American Bird Phenology Program, as the project is called, was created by the well-known Patuxent Wildlife Research Center. As the "about" section of Bird Phenology Program site describes, "The North American Bird Phenology Program houses a unique and largely forgotten collection of six million Migration Observer Cards that illuminate migration patterns and population status of birds in North America. These handwritten cards contain almost all of what was known of bird distribution and natural history from the Second World War back to the later part of the 19th century" (U.S. Geological Survey 2008). Although this mass migration observation project involved more than three thousand participants watching birds and keeping notes on them, it petered out in the 1970s. With new birdwatching technologies, we may be entering a new phase of mass *automated* observation as a form of technocultural birdwatching.

In technobirding, field guides will still have a role, but other tools will be wired, networked, visual, and acoustic. Many of these tools are available already, and they are only getting more integrated into birdwatching and daily life. In the World Series of Birding, one of three competitive birding practices examined in the next chapter, social software tools such as live Twitter feeds are increasingly being used by team members to report sightings (Chu 2009). Such competitive, sometimes high-tech birding practices take place in an array of polluted environmental areas, however, suggesting that the guides birdwatchers use to identify birds do not always facilitate what can be termed "green" environmental practices.

CHAPTER FIVE

Birding on Toxic Land

Some of the natural areas where birders watch birds are toxic. The competitive birding event called the World Series of Birding, North America's most publicized big-day birding event, takes place in New Jersey, the state with the highest density and number of EPA Superfund sites in the nation (Environmental Protection Agency 2007a, 2007c). Another competitive form of birding, known as big-year birding, regularly involves participants looking for rare birds at active trash dumps such as the Brownsville dump in south Texas (Kaufman 2000, 94–104; Obmascik 2004, 120–23) and reclaimed dumps such as the Montlake Landfill in Seattle. Similarly, birders involved in the practice known as listing sometimes look for birds at sewage treatment facilities, many of which process toxic sludge into fertilizer. The environment and science journalist Jeff Tollefson reports that "some 60% of the residual sludges from the process—several million dry tonnes annually—are now used as fertilizers rather than being buried or incinerated" (Tollefson 2008, par. 7).

In this chapter, I supplement the previous discussions of field guides and birdwatching to attend to representations of the more task-oriented and sometimes competitive bird-identification activities of *birding*. Studies of other organized outdoor pastimes have shown that daily practices are embedded with political and environmental perspectives. Iain

Borden has argued, for instance, that skateboarding is not just a hobby but a set of practices that redefine aspects of urban and suburban environments (2001). In a study of parkour, in which participants interact acrobatically with such things as stairwells, rooftops, and parking structures, Michael Atkinson describes practitioners as having aesthetic and spiritual commitments while "challenging dominant social constructions of [the] urban environment" (Atkinson 2009, 179). For Atkinson, parkour is not just a gymnastic overtaking of urban spaces; it is an anarcho-environmentalist practice that challenges fundamental relationships between the body, mind, city, and "nature" (170–71).

Birding is much less radical. In particular, the practice of birding in toxic locations creates a text that can be read as environmentally conservative given the ways it validates ongoing and entrenched forms of environmental degradation. When birders frequent the sites I discuss, they create readable texts of their practices that enact a rhetorical inversion what Kevin DeLuca has called an image event. For DeLuca, "tactical image events" are vivid scenes of ideological conflict, protest, and confrontation, and they have been used by environmental activists to dislodge dominant environmental ideologies in the public sphere (DeLuca 1999, 2–22). DeLuca writes that "radical environmental groups are using image events to attempt both to deconstruct and articulate identities, ideologies, consciousness, communities, publics, and cultures in our modern industrial civilization" (17).

By birding at toxic sites, particularly in widely publicized birding competitions, the practices of competitive birders sanction mainstream environmental ideologies. The presence of birders at such sites creates a sense that far from having no human utility, polluted environments provide a service to niche groups of environmental enthusiasts. Through these practices, I argue, a subset of the larger birdwatching community creates a risky sense that environmental toxins have become harmlessly integrated into what are imagined as thriving industrial ecosystems.

Even while some competitive birders take advantage of toxic sites, competitive birding gets billed as environmentalist. Birding is like birdwatching in having benefited historically from conservationist policies, habitat protection, and environmentalism more broadly. For instance, federal legislation to protect birds such as the Migratory Bird Treaty Act of 1918 and the Endangered Species Act of 1973[1] have limited overhunting and human impact to several endangered bird species. The establishment of wildlife refuges and bird sanctuaries has helped sustain breeding bird populations in particular (Philippon 2004, 72–105). Radical environmental groups such as Earth First! have used direct action to produce image events (DeLuca 1999, 104, 159) meant to intervene in the logging of old-growth forests and thus help preserve such threatened species as the Spotted Owl (Moore 1993).

Although these forms of environmental action have helped protect birds, thus benefiting birders, carefree birding at toxic sites makes the pressing environmental problem of toxic pollution seem benign. Some competitive birding events do raise funds for environmental advocacy groups and many birders never visit toxic locales, but constructing birding at toxic sites as part of conservationism or environmentalism overlooks the anti–image event the practice creates.

Phaedra Pezzullo, in her book *Toxic Tourism: Rhetorics of Pollution, Travel, and Environmental Justice* (2007), describes a form of environmental activism in which participants tour polluted sites in a very different way, bringing attention to problems stemming from toxicity. Birders tour such toxic sites as well, but their tours are anything but the toxic tours Pezzullo describes. To add a new species of bird to one's list of species at a toxic site is to indirectly revel in the presence of toxicity. Birds and birdwatchers inhabiting a landfill or sewage treatment facility can create a sort of text that makes it seem as if contaminants are not really a problem.

In addition to developing this discussion about birding and toxicity, this chapter defines a class of outdoor activities I call environmental sporting. After describing how a number of sports fit this profile, I focus on a case study of competitive birding. As I will describe, there is a fractious relationship between environmental sporting and environmental conservation such that birding depends on environmental degradation and conservation simultaneously, something that is bound up in the ideology of this kind of outdoor activity. As the examples show, some of the birds that competitive birders seek live amid toxins that travel readily into groundwater, lakes, streams, and food chains, yet in the inverted image event produced by acts of birding at toxic sites, such dangers are presented as a necessary part of the game birders play.

BIRDING AS A SPORT

Recall from the introduction to this book that birding includes many of the features of birdwatching—nature appreciation, aestheticism, and scientific classification—but that birders are also listers. Listing, as it is called, is the practice of keeping careful track of each species one identifies in the field over a set period of time, in a particular region, or both.[2] Birders tend to be more eager to find rare birds than birdwatchers are and seldom wait for birds to come to them. Instead, birders head out in search of yet-unseen species to add them to their lists. Birding becomes blatantly competitive at listing events such as the World Series of Birding.

In sporting events of this kind (the Montezuma Muckrace and the Great Texas Birding Classic are two other examples), teams of birders compete, receive prizes, and even challenge themselves physically. Corporate sponsorship funds some of the larger competitive birding events much like in mainstream sports. Listing competitions among birders can be intense, with participants birding nonstop for twenty-four hours or several days, taking few if any breaks, and working exhaustively to identify every bird possible.

Based on popular perceptions of birdwatchers as folks who stand in fields or parks watching birds through binoculars, it may be hard at first to accept that birding is a sport. How can a person studying a Yellow-headed Blackbird or sorting through species of warblers be participating in a sport? This reluctance to see birding as a sport results from a narrow definition of sport itself and a lack of understanding of all that goes into birding and twitching. (Twitching is the bird-identification activity that involves the spontaneous, fervent seeking of rare birds.)

To address the first impediment to seeing birding as a sport, note that many North Americans tend to associate sport with football, basketball, baseball, and other sports one can watch in a stadium or see in large-scale sporting events such as the Olympic games. Indeed, some textbooks provide equally narrow definitions of sport, calling sports "institutionalized competitive activities that involve rigorous physical exertion or the use of relatively complex physical skills by participants motivated by internal and external rewards" (Coakley 2006, 11). Other authors define sport much more broadly. Michael Mandelbaum, for instance, compares sport to religion, suggesting that sport can be categorized as those things that offer "a welcome diversion from the routines of daily life; a model of coherence and clarity; and heroic examples to admire and emulate" (2004, 4).

Rather than debate the narrow and broad definitions, I follow Kenneth Sheard (1999) and Raymond Korpi (1999), who argue that competitive birding has many aspects of even the most mainstream sports. Birding can be physically grueling, involve competing teams, and include awards for the winners of competitive birding events. Unlike many mainstream sports, though, competitive birding does not take place in the controlled environment of a field or stadium; instead, it is like other environmental sports in taking "nature" as its playing field.

ENVIRONMENTAL SPORTING

The sport of birding can be grouped into a category I call environmental sporting. Unlike other forms of sport, environmental sporting is not largely defined by human actors, but instead relies on human interaction with animated and fixed elements in what is thought to be a natural[3] or sustainable[4] environment. In fact, natural actors and environmental features have large roles to play in how environmental sporting is accomplished.

Just as the running river and blowing wind are agents in river kayaking and windsurfing, respectively, so are bird migrations and bird characteristics the crux of competitive birding. Some of the most obvious instances of environmental sporting include sports such as rock climbing, kayaking, canoeing, hiking, skiing, mountaineering, open-water swimming, outdoor cycling, and outdoor running. These sports have certain environmental requisites (for example, high river volumes, intact glaciers, well-maintained trails) that make each sport possible, and the enjoyment of these activities is in part derived from direct interaction between the participant and an environment. Birders have an ongoing and pressing need for a wide variety of living bird species and adequate depth to those bird populations, so species degradation and extinction conflict with the needs of birders just as eroded trails and dammed rivers conflict with the needs of hikers and river kayakers.

Like the other environmental practices mentioned—parkour and skateboarding—environmental sporting has ideological and political components. Perhaps the best example is adventure racing, which was made famous by Mark Burnett's EcoChallenge races of 1995–2002. In adventure racing, individuals or teams set out to traverse various forms of "wilderness," fording streams, rappelling down cliff faces, and climbing mountains.

Barbara Barnes describes adventure racing of this kind and the EcoChallenge race in particular as a "strikingly unapologetic and uncritical valorization of colonial exploration" (2009, 238), in part because the EcoChallenge races were held in places such as Borneo, Australia, and southern Utah. In this reading, adventure races become texts that, when viewed, reproduce dominant ideologies. It does not matter if the stated beliefs of participants and sponsoring organizations are green or if the events raise money for environmental conservation—environmental sporting does other kinds of cultural work as well.

In using the word *environmental* in environmental sporting, it is important to note that environmental sporting is not necessarily environmental*ist* in that such sports only infrequently involve advocacy for environmental conservation and protection. Nonetheless, environmental sporting is generally reliant on the ongoing success of environmental conservation (via new forms of legislation, land preservation, and activism) that protects the natural resources that environmental sporting makes use of. Because of this indirect relationship to conservation and environmentalism, in recent decades environmental sporting has become part of what has been referred to as the greening of sport (Maguire 2002, 91–92). Green sports are characterized as having an association between participation in the sport and green environmental politics, even if such a relationship is indirect or superficial.

Many forms of environmental sporting damage landscapes and rely on the consumption of an array of products that in many cases damage and pollute more landscapes,[5] but engaging in environmental sporting is imagined, through the greening of sport, to contribute to environmental protection. My purpose is to question this logic, but I also note that connections between birding and environmental protection are not entirely unfounded. Many arguments and initiatives for environmental preservation and conservation have cited the needs of sport practitioners. Such was the case with early lobbying efforts by hunters in the Boone and Crockett Club that resulted in the protection of national forest preserves and structured hunting laws (Philippon 2004, 52–53); similarly, the needs of hikers are routinely cited by habitat-preservation groups such as The Nature Conservancy (n.d.).

So, even though this kind of sport may not directly contribute to environmental protection, environmental sporting relies on environmental conservation while conservation relies in part on the sport. Such mutual reliance is integral to environmental sporting. Although the relationship between environmental sporting and conservation has been symbiotic, one aim of this chapter is to question the idea that environmental sporting promotes what could be called an environmental ethic.

THE BRIEF HISTORY OF BIRDING AS ENVIRONMENTAL SPORTING

As pointed out at various points throughout this book, the need for healthy and thriving bird populations has been anything but an abstraction in the history of birdwatching. In fact, it was because of a direct threat to many species of wild birds in the 1880s and '90s that bird-identification activities were first widely sponsored in North America by Audubon societies (Graham and Buchheister 1992), the institution of national Bird Day in 1894 (Weidensaul 2007, 155), and the publication of some of the first bird-identification field guides (described in chapter 1).

Conservation efforts aimed at saving birds at the end of the nineteenth century were multipronged in that they involved sponsoring protective legislation, establishing bird reserves and refuges (Gibbons and Strom 1988, 136–37), and trying to undermine the popularity of fashions using bird feathers (Price 1999, 57–109). In the 1880s and '90s, sponsoring the pastime of birdwatching was an additional way that new, favorable attitudes toward birds and the environment could be produced and maintained.

Note, however, that I did not say that was the case with the environmental sporting practice of birding. Instead, it was the hobby/pastime of bird*watching* that was most popular in the late nineteenth century, with the sport practices of birding and twitching taking shape by the 1950s, the era Steven Gelber (1999) describes as the heyday of hobbies in the United States. But in the way that competitive birding incorporates and is built on the central concerns of birdwatching (nature appreciation, aestheticism, and scientific classification), birding also owes its existence as a form of environmental sporting to the many responses to threatened bird populations.

With a focus on listing species, birding as we now know it took off in the 1950s and '60s, based on the field-mark identification system of Roger Tory Peterson's *A Field Guide to the Birds* (first published in 1934) and the support of the American Birding Association, established in 1969 (Weidensaul 2007, 257–58). As listing grew in popularity, it remained dependent on the success of ongoing conservation efforts to protect individual species such as the Peregrine Falcon, Bachman's Warbler, and Spotted Owl. Without these birds, lists would shrink. To attempt to protect these birds, particular environmental toxins (DDT) and habitats (wetlands and old-growth forests) have been the targets of advocacy.

The most obvious recent demonstration of this reliance of birding on environmental conservation is in relation to the 2004 rediscovery[6] of the Ivory-billed Woodpecker, a species long thought to be extinct. News of the rediscovery was made public in 2005 and marked by two events: an announcement that eighteen thousand additional acres of the species' habitat had been purchased by The Nature Conservancy to protect the remaining birds (Fitzpatrick 2005) and the online publication by field-guide author David Allen Sibley of a new insert for his *Sibley Guide to Birds* describing the Ivory-billed Woodpecker. Now birders had one more species to look for.

In this example and others, conservation and birding go hand in hand in the U.S. context: the more living species of birds, the longer birders' lists can be. It is no coincidence that every major field guide to birds currently in print in the United States is sponsored by a major environmental nonprofit organization, a mutually beneficial relationship.[7]

Mainstream environmental organizations such as the National Audubon Society and National Geographic Society have associated themselves with the cause of bird conservation; in return, these groups benefit from the association and profit from the sale of the field guides they endorse. Environmentalist groups with more radical and controversial agendas such as Earth First!, People for the Ethical Treatment of Animals, and the Earth Liberation Front remain unconnected to and unencumbered by the moderate environmental politics of birding. The brand of environmentalism promoted by mainstream environmental organizations fits nicely with the sport practice of birding. Although calls to end global warming threaten to radically change entrenched aspects of industrial capitalism, calls to protect wild birds have thus far only involved relatively small changes: the establishment of trade and hunting laws, small-scale nature preserves, and pesticide regulation. The forms of environmental protection that have been thought to benefit birds and sustain birding do not threaten to unsettle larger socioeconomic systems; instead, the environmental politics of birding fit within that framework.

What is unusual about birding, as discussed in the following sections, is that the specter of environmental degradation is not always avoided by birders in the ways it is in other "green" environmental sports. Whereas hikers, for instance, generally tend to prefer hiking in what seem to the hiker to be unadulterated landscapes and river kayakers generally prefer to run clean (or seemingly clean) rivers, some birders seek out polluted environmental niches. Toxic, polluted environments are frequently where the birds are and, as a result, where some birders are. There is a certain kind of black humor in birding at a toxic site, and because of the relationship between birding and toxicity, birding often functions as a form of environmental sporting that brings environmental pollution immediately into, not out of, view, even while making light of the seriousness of that pollution.

TOXIC ENCOUNTER WITH BIRDS NO. 1: BIG-DAY BIRDING AND SUPERFUND SITES

The most well-known competitive birding event in North America, called the World Series of Birding, is an annual competition run by the New Jersey Audubon Society that attracts teams of expert birders and corporate sponsorship. This annual event, begun in 1984, consists of "a 'competitive' Big Day" in which "you have 24 hours to identify as many species by sight or sound. Each species seen or heard counts as one. The playing field is the state of New Jersey" (New Jersey Audubon Society 2007).

In some ways, the World Series of Birding is akin to the EcoChallenge adventure races mentioned above, only the adventure involves the search for birds and motorized vehicles are allowed. Participants can work alone or in groups, there are separate categories for youths and seniors, and although there are options to travel the state looking for birds or stay fixed in a single location, the traveling, statewide competition is the most prestigious. In this event, competitive birders pile into cars[8] in the early hours of the morning, scour the state in search of birds, and often finish at midnight, listening in the dark for owls and other nocturnal birds. Some species are spotted out of open car windows, others in parking lots, and many in parks or along bodies of water.

Self-described as "North America's premier conservation event," the World Series of Birding raises money (through entry fees, pledges relating to the number of species identified, and corporate sponsorship) for the New Jersey Audubon Society and an array of conservation efforts. Prominently displayed on the website for the event is the statistic that, as of 2007, the World Series of Birding had "raised over $8,000,000 for bird conservation" (New Jersey Audubon Society 2007).

Although funding conservation is a clear purpose and accomplishment of the event, the World Series of Birding also sponsors competitive birding more broadly, raises awareness for what can be the plight of migratory birds, and supports stakeholders in North American birding and conservation: the New Jersey Audubon Society, Cornell University's Lab of Ornithology, sponsoring optics manufacturers, and sponsoring utility companies. The event's emphasis on migratory birds makes sense for several reasons: New Jersey is on the eastern migratory flyway (a major thoroughfare for migrating birds), the World Series of Birding is held at the height of spring migration, and the event coincides with International Migratory Bird Day.[9]

The New Jersey Audubon Society describes the World Series of Birding as "focus[ing] national media attention on the challenge and adventure of birding" (2007), and media outlets such as National Public Radio (Solomon 2003) and *The Daily Show* (2000) have run stories on the event. What remains unpublicized, however, is the way the participants in the World Series of Birding both coexist with and take advantage of polluted landscapes within New Jersey. The World Series of Birding is positioned as a green sporting event that raises money for conservation and awareness about migratory birds, but at the same time is a competitive environmental sporting event that thrives in part because of toxic landscapes, with some key sightings each year taking place at landfills and other toxic areas (Binns 2003; Cornell Lab of Ornithology 2007).

In the World Series of Birding, the context of team competition and the pressure of the clock lead to a diminishment of what could be called an environmental conservation ethic. Such an ethic might focus directly on remediating the toxicity of New Jersey's more polluted habitats; instead, the event prioritizes raising money "for bird conservation" as the best, most effective, and only way to sponsor environmental conservation.

The tactics of in-your-face environmental activism practiced by Earth First!, Greenpeace (DeLuca 1999), and some organizers of performance-oriented toxic tours (Pezzullo 2007), for instance, do not surface in this conservative conservation event. Instead of protest, culture jamming, confrontation, and direct action, the World Series of Birding focuses on competitive fund-raising.

Although New Jersey often gets caricatured as a polluted wasteland, the Garden State is as amenable as any other for outdoor enthusiasts. Birders like it in particular because it is on the eastern flyway and has an extensive shoreline. With that said, the state does have its share of environmental problems. Having been home to a wide range of industries, New Jersey now has the largest number of EPA Superfund cleanup sites[10] of any state in the nation even though it is the fifth smallest state. Superfund sites are properties and areas that have been identified by the EPA as toxic enough to require immediate government intervention; the descriptions of Superfund sites in New Jersey characterize most of them as abandoned or mismanaged properties where a range of contaminants, often in powerful combination, were dumped into the soil and water.

The World Series of Birding may be "a heck of a lot of fun," as the New Jersey Audubon Society describes the event on its website, but part of that fun is derived from a direct interaction with these toxic parts of New Jersey. A participant, from 2007, describes one of his teammate's actions in this way: "Brian made some amazing picks—like the Iceland Gull high atop the landfill 'mountain' across the river at Florence. That gave us four key birds to pay for our difficult decision to route in Florence, and a huge morale boost during the long trek to the south near midday" (Cornell Lab of Ornithology 2007). The avian bounty provided by trash "mountains," in this account, boost the morale of the team as they become absorbed in the thrill of the sport. There is a kind of thrill, in such accounts, in identifying birds amid the filth of a landfill or other toxic location.

Also in Florence, New Jersey, the city where this Iceland Gull was identified, is a site once occupied by Roebling Steel Company, now EPA Superfund site number NJD073732257. In a fact sheet about the site issued by the EPA, the property is described in this way:

> The site includes two inactive sludge lagoons, an abandoned landfill, buildings containing pits and sumps, contaminated soils and slag material, contaminated river and creek sediments, impacted wetlands, and contaminated groundwater. . . . Buildings on the site contained contaminated process dust and exposed asbestos. Ground water under the site is contaminated with various heavy metals including chromium, lead, cadmium, nickel, zinc, and copper. Soil all around the site is contaminated with heavy metals, including lead. River and creek sediments are contaminated with heavy metals and polycyclic aromatic hydrocarbons. People on-site could come into direct contact with hazardous materials or could accidentally inhale contaminants from the soil and process dust in the buildings. Runoff from precipitation on the site may have contaminated the Delaware River, which is next to the site. (Environmental Protection Agency 2007b)

Although I have no evidence that the Roebling Steel Company produced steel used in the binoculars or spotting scopes birders use, the EPA reminds us that heavy metals such as lead and cadmium, employed in the production of less environmentally friendly glass used in binoculars, are present at the site.

Polluted sites of this kind can be directly hazardous to birds, as was the case in 1998 and 1999 when five hundred to one thousand birds died within a few days at Florida's Lake Apopka due to toxic levels of pesticide residues (Patterson 1999; Environmental News Service 2003). Similarly, toxic PCBs from polluted sites have been found to make their way into the bodies of birds, causing such symptoms as heart deformities and higher mortality rates (Indiana University 2006). But the threat of toxics to birds is just a small part of the larger environmental picture. Toxins at EPA Superfund sites such as the Roebling Steel Company property leach into groundwater and get distributed throughout multiple food chains.

Finding rare gulls at the landfill in Florence, New Jersey, is nothing new to birders at the World Series of Birding. In fact, the same species (*Larus glaucoides*) that was sighted in 2007 was identified there by the winning team in 2003. This earlier sighting was described in this way: "Swinging into Florence, Great Cormorants are on the near marker and Lesser Black-backed Gull and a 2nd year Glaucous Gull on the Delaware [River]. It takes a little while but eventually an Iceland Gull shows itself on the landfill" (Binns 2003). This stretch of the Delaware River runs alongside what was Roebling Steel and is now an EPA Superfund site. In fact, the major danger posed by this site is its proximity to the river, as groundwater is only ten feet below the ground and is readily leaching contaminates into the river and water table (Environmental Protection Agency 2007b).

Florence is by no means the only polluted part of New Jersey that competitive birders frequent in their quests to tally a maximum number of birds in "North America's premier conservation event." Posting to an online forum from 2003, a participant asks "Where'd you get your Kingfisher?" referring to a relatively common species, the Belted Kingfisher. The participant then notes: "Our one and only was at the Sussex Landfill" (Bernzweig 2003). It seems almost as easy to find an EPA Superfund site in New Jersey as it is to locate a new species of bird: there are eight Superfund sites in Sussex County, New Jersey, each with a profile similar to or worse than that of Roebling Steel Company.[11]

Above, I said that fast-paced competition can override what might be an environmental ethic more attuned to the sometimes toxic landscapes of New Jersey. In several ways, the blinding forces of competition may be to blame for causing participants to overlook environmental conditions of toxicity and pollution. When searching for bird species in a finite amount of time while trying to win a competitive birding event such as the World Series of Birding, it may become easier for participants to overlook suspect environmental conditions that affect birds and other living things. I am more inclined to fault the structure of the event than the participants, however, because the rules for the World Series of Birding are such that no region is off-limits within the state. By defining the entire state of New Jersey as the playing field for the event, the World Series of Birding promotes the troubling notion that birds and other wildlife thrive equally well in protected ecosystems as they do in heavily industrialized ones.

The World Series of Birding and New Jersey Audubon are also enmeshed in various industries through corporate sponsorship. The list of corporate sponsors for the event includes the nuclear power plant company AmerGen Energy, several other utility companies, and a full range of binocular makers, including Bushnell, Zeiss, Leica, Leupold and Stevens, Nikon, Pentax, Steiner, and Swarovski. The relationship between utility companies and environmental conservation is fraught with challenges, and binocular makers have a history of making glass containing such pollutants as cadmium, lead, and arsenic. Sponsoring such "environmental" events is part of a strategic public-relations strategy that utility companies engage in to position their industries as somehow green.[12]

Because of the corporate funding for the World Series of Birding, the event sustains its own problematic relationship between birding and environmental conservation: the World Series of Birding aims to sponsor bird conservation while celebrating the coexistence of birds and toxic pollution. As a result, the seemingly healthful and vibrant nature of environmental sporting of this kind ends up inoculating outside viewers of the event that is the World Series of Birding to the real perils of toxic pollution.

As Joseph Maguire has shown in his discussion of what were initially slated to be the most "green" Olympic games, the Summer 2000 Olympics in Sydney, even when efforts have been made to make large-scale sporting events less environmentally detrimental, such changes have not always been realized. Maguire concludes, "Commercial imperatives dominating the culture of sport can often override practices that may protect the environment and human health" (Maguire 2002, 91). The utility companies that sponsor the World Series of Birding clearly do so to green their images; the binocular companies sponsor the event to raise their profiles and sell more product. The type of competitive birding manifest in events such as the World Series of Birding relies directly on and indirectly contributes to the degradation of the very landscape the birds of New Jersey migrate to and through.

TOXIC ENCOUNTER WITH BIRDS NO. 2: BIG-YEAR BIRDING AT LANDFILLS

Big-day birding of the kind found at the World Series of Birding demonstrates how participating in an environmental sporting event can appear to be a major way to advocate for "the environment," although such participation also stands in for alternative forms of environmental action that might do more to challenge pressing environmental problems such as toxicity and pollution. Not all competitive birding is as centralized, organized, and as heavily sponsored as the World Series of Birding, though.

In what is known as big-year birding, birders dedicate a year to identifying as many species of birds as possible. Since at least the late 1960s, birders seeking to set the big-year record have traveled North America, making regular stops at landfills. In big-year birding at landfills, we see a kind of divestiture from traditional environmental commitments to the conservation of land, water, and air due to the demands of listing a record number of species in what is called a "big year."

Big-year birding is a big deal in the birding community, with a steady stream of books being published on the topic and celebrity status passing from each reigning big-year champion to the next.[13] Big-year birding is defined as the competition to identify the greatest number of birds in a bounded region in a single calendar year, and most publicized big years, as this kind of birding is called, are generally bounded by the national borders of the United States and Canada (excluding Hawaii). Within artificial boundaries and amid frequent changes to the number of species designated by the American Ornithologists' Union (AOU) (Karnicky 2007), big-year birders compete to see who can break the record and identify the greatest number of species in a year.

Big-year birding is an outgrowth of 1950s birding culture in several ways. Although what is generally considered to be the first big-year effort was conducted by New York banker Guy Emerson in 1939 (Weidensaul 2007, 289), big-year birding was first popularized on a grand scale by field-guide author Roger Tory Peterson in 1953, a year when Peterson set the record by identifying 572 species in North America. Much of this adventure by car and plane was described in Peterson and James Fisher's book *Wild America* (subtitle: "The Record of a 30,000-Mile Journey around the Continent by a Distinguished Naturalist and His British Colleague") (1955). Although Peterson did not dedicate the entire year to birding in the way that later big-year birders would, he and Fisher scoured most corners of North America in search of birds. Their book is compelling and suspenseful and reads like adventure writing. With the publication of *Wild America*, big-year birding was born in a spirit of 1950s automobile and hobby cultures (Gelber 1999).

The obvious contradiction within big-year birding is that it has typically been practiced in ways that involve large expenditures of fossil fuels for transportation, and in having a considerable carbon footprint, big-year birding is similar to the World Series of Birding. This problem does not stand out in all accounts and instances of big-year birding, however. Kaufman's big year, which took place in 1972, involved Kaufman traveling largely by hitchhiking.

Even more in tune with the connection between burning fossil fuels and environmental degradation, in 2007 the Boothroyd family set out on a self-publicized big-year birding expedition across North America—traveling entirely by bicycle. Calling their adventure a "bird year" instead of a "big year" to distinguish a new form of environmentally sensitive big-year birding, this bicycling-and-birding adventure functioned as a deliberate critique of the wastefulness and exorbitant cost of more traditional, 1950s-style big-year birding (Boothroyd, Boothroyd, and Boothroyd 2007–2008).[14]

As big-year birders such as the bicycling Boothroyd family address environmental concerns associated with automobile transportation and big-year birding, I focus on the implications of landfills and trash dumps as they have become permanent fixtures of much big-year birding. As with the World Series of Birding, where participants sometimes visit landfills and other toxic sites to add hard-to-find birds to their lists, big-year birding has developed in such a way as to make certain well-known and heavily publicized landfills regular stopping points on a big-year birding odyssey.

Landfills are popular with big-year birders because they are easily accessible and because scavenging species of birds thrive there, feeding on food scraps available on the surface. Beneath landfills, however, is where toxins can spread into groundwater via moving leachate. Two well-known birding landfills—the Brownsville dump in south Texas and the Montlake Landfill in Seattle—highlight how the toxicity of such sites is not only overlooked or ignored by birders as part of environmental sporting, but relied on and even valued by practitioners. In a sense, toxic landscapes such as the Brownsville dump and Montlake Landfill are critical aspects of the polluted playing field of this kind of competitive birding.

In his account of big-year birding in the early 1970s, Kaufman includes a chapter he titles "To the Promised Landfill." He writes: "Sometime in the 1960s, apparently, the crows wandered north and discovered that there was food in the huge dump at Matamoros, just south of the rio. From there it was only a short hop across to the Brownsville dump on Boca Chica Drive, where the pickings were richer, because American threw away more food than the Mexicans did" (Kaufman 1997, 97). *Corvus imparatus*—formerly called the Mexican Crow but renamed, by the AOU in 1997, the Tamaulipas Crow—remains a highly sought-after bird when it comes to big-year birding (Scott 1999–2002; Obmascik 2004). Although most nonbirders think that all crows are alike, big-year birders are quite particular about listing each species of crow found in North America.

With the AOU currently listing nine species of crow as possible finds in North America (Banks et al. 2007), somewhere between four and nine species of crow can reasonably be expected to show up on a big-year birder's list. The prospects of adding one more crow to the list has long brought birders to the Brownsville dump. In his account of big-year birding, Obmascik describes birders arriving at the Brownsville dump by the carload, being routed to a special viewing area, and then forming "a neat line of scopes and tripods" with "everyone seem[ing] to be breathing through the mouth" (Obmascik 2004, 122) due to the stench. Because of the volume of birders coming to the dump to see the Tamaulipas Crow, a special viewing area had to be set aside amid the garbage.

To understand how the Brownsville dump functions in the ideology of birding and big-year birding, it is crucial to understand that Tamaulipas Crows have not historically been hard to find. In fact, the crows have been relatively abundant in a relatively small part of northeastern Mexico. The rules of big-year birding stipulate, however, that it is necessary to identify a species within a certain bounded geographic area, and the Brownsville dump has been one of the only places *within the contiguous forty-eight states* that is both readily accessible by car and frequented by the Tamaulipas Crow.

The food scraps at the dump, in a sense, have served as the attractor for this species of North American crow, and along the borderlands of south Texas, it is the Brownsville dump where big-year birders have historically gone to add Tamaulipas Crows to their lists. "Nobody besides the crow liked going there," Obmascik writes about the dump. "To say it stunk did injustice to the word *stunk*. It reeked. It rotted. It marinated decades of throwaway table scraps in the fecund humidity of the Rio Grande Valley and then roasted it under the South Texas sun" (Obmascik 2004, 121, emphasis in the original).

In recent years, however, the populations of Tamaulipas Crows have vanished from the dump (Scott 1999–2002). Carloads of birders once drove into the dump to add this species of crow to their list, but the numbers of Tamaulipas Crows seem to be decreasing, their range is changing, or both are occurring. In the case of this species of bird, birding in general and big-year birding in particular relied on this dump in south Texas to attract a species of crow into the United States so that it could be listed. The polluted site of the dump served as a playground for birders who could stand the smell (Obmascik 2004, 122), a fabled destination mentioned in guides to birding hotspots (Zimmer 2000, 240), and a kind of guaranteed border-opportunity for sighting this common-looking bird. But for the Tamaulipas Crows in residence at the dump, the trash from El Norte ultimately provided insufficient refuge. In a socioenvironmental climate in which crows have long been besieged by hunters who see them as pests, an attitude maintained by contemporary anti-crow-hunting groups such as Crowbusters.com, preserving crows, Tamaulipas or otherwise, was never integral to the practices of big-year birders. What mattered was spotting the species, listing it, and moving on.

Another landfill, the Montlake Landfill on the banks of Lake Washington in Seattle, bears some relationship to the Brownsville dump. The difference is that the Montlake Landfill no longer accepts garbage. The landfill was active for both burning and burying waste from 1926 to 1966 (Montlake Landfill Work Group 1999, 10–11), but was finally "capped with two feet of clean soil" between 1969 and 1971 (Montlake Landfill Oversight Committee 2002, 11). Since that time, attempts have been made to mask the history of the Montlake Landfill with the veneer of a more natural-sounding name of the Union Bay Natural Area (Center for Urban Horticulture, n.d.), but the original name of Montlake Landfill (or simply "the Fill") has remained. What was once an active waste repository now looks like any other city park: the cap of topsoil is landscaped with grasses, bushes, trees, walking trails, and small ponds.

That the Montlake Landfill borders on Lake Washington is significant for two reasons. First, it makes the Montlake Landfill a good place to find a wide range of bird species, with land, marsh, and water habitats at the site. The Montlake Landfill attracts birds and birders for this reason. Second, the Montlake Landfill poses significant dangers to Lake Washington. Damage to the lake in the form of lateral peat movement (Montlake Landfill Work Group 1999, 13) and toxic leachate, for instance, has surfaced as an ongoing concern in the two major reports about the site (the Montlake Landfill Work Group's "Montlake Landfill Information Summary" [1999] and the Montlake Landfill Oversight Committee's "Operational Guidance for Maintenance and Development Practices Over the Montlake Landfill" [2002]). As the "Montlake Landfill Information Summary" states:

> Municipal solid waste landfills generate leachate from processes of decomposition and water percolation through the waste. Most of the water entering the Montlake landfill is through ground water and infiltration of rain and surface water runoff. The mass of garbage and debris stored in a municipal solid waste landfill represents a finite source of pollutants. Most of the water-soluble pollutants leach out of the landfill through successive volumes of water, their concentrations diminishing over time. Other materials remain in the landfill for many reasons, including but not limited to absorption, water solubility, and particle size. (Montlake Landfill Work Group 1999, 16)

Although the lateral movement of compacted peat into the lake, which was caused by the daily compacting of garbage, was partially rectified in the 1950s with the construction of several underwater wooden containment dikes (Montlake Landfill Work Group 1999, 13), the leachate is not as easily contained or quantified.

Undetermined substances were dumped at the site while the landfill was in use, including what has been described as a "black acid concentrate" dumped by the Seattle Gas Company circa 1955 (Montlake Landfill Work Group 1999, 11). Montlake Landfill is not, however, the most toxic landfill in Washington State, and the Washington State Department of Ecology has determined the site to be relatively stable (Montlake Landfill Oversight Committee 2002, 6). In its current configuration, in accordance with the Model Toxics Control Act (WAC 173-340), the Washington Department of Ecology has determined that "if the Montlake Landfill is left undisturbed, there is a low risk of adverse impact to human health and the environment, and no remedial cleanup actions will be required in the near future" (Montlake Landfill Oversight Committee 2002, 6). This conclusion comes despite "asbestos material" having been encountered at the landfill and speculation that "asbestos-containing material may also be encountered at other locations within the limits of the Montlake landfill" (Montlake Landfill Oversight Committee 2002, 16).

There is also constant off-gassing of methane at the landfill as a result of garbage decomposition, and even though methane is relatively harmless in low concentrations, there is a risk of the entire landfill being destabilized in one of the earthquakes that sometimes shake the region. This risk is noted in the 2002 report, which asserts that "the Montlake Landfill is a critical area as defined by the City of Seattle Department of Construction and Land Use (DCLU) for liquefaction and methane mitigation" (Montlake Landfill Oversight Committee 2002, 11). Liquefaction refers to what would happen if a significant earthquake (of the kind that shook the area in 2001) were to disturb the landfill material and the thin soil cap covering the decomposing garbage.

In a black-and-white photograph of the Montlake Landfill taken in 1955, during its heyday as an active trash collection site (figure 5.1), an enormous flock of gulls hovers over a tractor at the landfill. The photo is a reminder that the site has always sustained bird populations, although the presence of birds has signaled very different things. When the landfill was active, gulls circling overhead signified the presence of garbage, whereas today, the many species of birds on site (including gulls) make the Montlake Landfill seem healthy and safe.

Here, a lakeside marsh became a landfill and was then turned back into a "natural area." Today, the Montlake Landfill looks like any other park. Birders now frequent the area, standing atop many cubic tons of compacted garbage that have been hidden from view by a thin layer of topsoil and some clever landscaping. Big-year birders moving through the area may park their cars at the parking lot on site, perceiving the area

Figure 5.1. A bulldozer operator works beneath a cloud of gulls at the Montlake Landfill. *Seagulls at Montlake Landfill, Seattle, 1955.* Josef Scaylea photo, Museum of History & Industry, Seattle.

as only and always a "natural area" that is a good place to see local birds. In this way, garbage remains hidden thanks to the preoccupation with seeing birds and overlooking the toxic history of the site.

Unlike birding at the Brownsville dump, for instance, where a kind of celebration of birds amid trash takes place, here toxic pollution is either ignored or seen as a healthy habitat "filled" with birds. The presence of thriving bird populations at sites such as the Montlake Landfill makes the toxicity of a capped landfill seem nonexistent. In a sense, the birds and birders perform their own public relations work for a city harboring a toxic site such as the Montlake Landfill.

TOXIC ENCOUNTER WITH BIRDS NO. 3: LISTING AND SEWAGE SLUDGE

Thus far, I have focused on how competitive birding relates to environmental conservation within some of the more highly competitive aspects of birding as environmental sporting: big-day and big-year birding. Here, I move to the more everyday and widespread practice of listing.

Both big-day and big-year birders keep lists, but listing need not be part of the organized, competitive birding found in big-day or big-year events. The lengths of birders' life lists are sometimes compared online and reproduced in such places as *Birding*, a publication of the American Birding Association. Birder Phoebe Snetsinger is legendary for having seen more than 8,400 of the 10,000-plus species of birds on the planet (Snetsinger 2003). Listing is motivated by a drive to identify as many birds as possible, and birding handbooks help listers by providing details about birding hotspots. Among such hotspots, it is common to find reference to yet another highly polluted ecological niche: sewage ponds.[15] Joey Slinger's *Down and Dirty Birding* (1996), for instance, dedicates a page to birding at "sewage lagoons" (46), and Pete Dunne mentions birding at sewage ponds multiple times in *The Feather Quest: A North American Birder's Year* (Dunne 1992, 73, 79, 80, 81). In Dunne's account, birding at sewage ponds is both comical (81, 85) and essential to finding certain species of birds (80).

But no handbook is more singularly focused on this particular toxic encounter than William Tice's birding guide *A Birders Guide to the Sewage Ponds of Oregon* (1999). "There is no one else on the planet," Tice begins his introduction by stating, "that has a vested interest in sewage ponds like birders" (ii). True to its title, Tice's entire guide is dedicated to helping Oregon birders find mellifluent sewage ponds and the birds they attract. It is both the water and the sludge, blooming as both are with algae and oversized insects,[16] that bring the birds and the birders to sewage ponds (Zimmerling 2006). At sewage ponds, birders can peer through the fence, spotting birds wadding and swimming around in the pools of sewage.

In comparison to EPA Superfund sites and toxic landfills, sewage ponds may not seem like a problem. If sewage ponds only processed pure human waste and made it into fertilizer, they would function as they are intended to: as vital waste recycling centers. However, sewage sludge (the solid material that results from processing human sewage) has been found to include a broad array of heavy metals and toxic chemicals; a congressional report has even found "several cases of radioactive contamination [to] have occurred at sewage treatment plants" (Wells et al. 1994, 14). Anything that goes down a drain or toilet in a household, hospital, or factory can wind up in the sludge at a sewage treatment facility, and because a percentage of processed sludge is used as fertilizer, these toxins can then be redistributed onto food crops.

Sewage treatment facilities were originally intended as an environmental solution meant to end the practice of dumping raw sewage into steams, rivers, lakes, and other large bodies of water. Although officials at the EPA have attempted to rename sludge with the more benign-sounding term *biosolids* (Perciasepe 1996), many anti-sludge environmental advocacy groups (Vermont Public Interest Research Group 1999; Parnell 2001) and scientists (Snyder 2005; Harrison et al. 2006) have published detailed accounts of the problems associated with spreading toxic sludge on food crops and near communities.

As Caroline Snyder writes, the "EPA [has] forged a powerful alliance with municipalities that needed an inexpensive method of sludge disposal and sludge-management companies that profit from this practice. The alliance's primary purpose was to control the flow of scientific information, manipulate public opinion, and cover up problems, in order to convince an increasingly skeptical public that sludge farming is safe and beneficial" (Snyder 2005, 415). As these reports and research show, the use of sludge as a fertilizer can spread heavy metals, chemicals, and disease both back to humans and up various food chains.

So, even though sewage treatment facilities seem to represent a triumph over the dumping of raw sewage into open bodies of water, recycling sludge remains an unreconciled environmental problem. Birders, however, light-heartedly and enthusiastically expand their lists of birds at sewage ponds. Some of the more popular ponds even have permanent signage stating where birders can and cannot go. Much like competitive birding at Superfund sites and active landfills, birding at sewage ponds makes light of the toxic playing field of birding. The very presence of birders at sewage ponds makes sludge seem anything but the toxic environmental problem that it is, and birders only magnify this image.

Some toxic sites are too polluted for any human use other than birdwatching. Certainly there are few other recreational uses of active landfills and sewage ponds than birding. By promoting the public perception that toxic landscapes of this kind are useful for environmental sporting, though, birders who frequent these sites are in a sense endorsing ongoing environmental degradation on multiple levels. Earlier I referred to how Michael Atkinson interprets some who practice parkour as "anarcho-environmentalists" who challenge dominant environmental ideologies via their daily practices (Atkinson 2009). In contrast, the daily practices of some birders sanction such ideologies. In addition, the environmental problems associated with sludge are completely disregarded in the literature on birding at sewage ponds; instead, sewage ponds are described as benign (if smelly) places to find more birds. The practices of sewage-pond birders are in concert with EPA efforts to deny any problems associated with the repackaging of sludge as fertilizer.

BAGGING BIRDS

The forms of competitive birding discussed are their own image events that mask or make light of several pressing environmental problems. Although birdwatching has had a long history of associations with environmental conservation, such engagement is limited in these forms of competitive birding.

Such is the case even though, in the case of birding, environmental conservation is a central component of this "green" sport. Every new field guide to North American birds sold is affiliated with and raises money for one nonprofit environmental conservation group or another, and competitive big-day events such as the World Series of Birding contribute funds to aid environmental conservation. Like other environmental sporting practitioners, birders are frequently cited as stakeholders in efforts to help justify a range of environmental protection efforts. Because protecting environmental assets that benefit the interests of environmental sporting stakeholders is politically salient in comparison to protecting environmental assets with no tangible human benefit, sports such as birding have played a role in shaping mainstream environmentalism in the United States. Said another way, environmental sporting justifies moderate forms of environmental conservation on human terms, for human utility, not based on any intrinsic commitment to environmental protection in and of itself.

Although competitive birding has been described as helpful to conservation efforts in North America, it is particularly problematic that birding at and near toxic locations such as Superfund sites, landfills, and sewage ponds obscures a range of pressing environmental hazards. Birding has become a component of mainstream, conservative environmental protection efforts in the United States.

As Kevin DeLuca has argued, environmentalism runs the risk of becoming quickly coopted and thus reproductive of mainstream ideologies (DeLuca 1999, 71). That birds and birders appear to safely inhabit and enjoy toxic environments masks some of the real but less visible hazards associated with leaching toxins, in the case of Superfund sites and landfills, and recycled toxic sludge, in the case of treated municipal sewage. It is not that the pollutants at the sites discussed directly harm birds in every case, although such cases certainly do exist, but rather that birders create an inverted image event that makes it appear as if "nature" can thrive even amid toxics. This view renders toxicity on the whole as if it were just another benign human addition to an already altered world. As birders seek the next bird to add to their lists, they move through toxic environments as if they were sanitary and harmless playing fields.

Birding at toxic sites is a form of environmental sporting that accepts environmental toxicity as just another part of a seemingly healthy ecosystem populated with resilient birds and thus ultimately challenges some of the goals of conservation and preservation. Ultimately, though, an examination of competitive birding reveals that some of the more conservative forms of contemporary environmental conservationism, based largely on fund-raising efforts, are implicated in perpetuating ongoing aspects of environmental degradation.

Environmental sporting operates not despite, but through oversights within environmental thinking. In its most competitive forms, environmental sporting is about focusing on the summit, the rapids, the trail, the next big wave, the EcoChallenge, or the rare species of bird while overlooking potential costs associated with these activities. The most obvious costs—added greenhouse gases and environmental pollution—are part of all environmental sports; it could be called the Mount Everest effect, named for the vast quantities of expeditionary waste that have accumulated on that famous peak. More important, though, is that environmental sporting does ideological work, constructing environmental sporting as conservationist while diminishing the perceived problems associated with toxic pollutants in the sporting playing field. Engaging in sport at toxic sites overwrites those sites as healthy.

CONCLUSION

The Birdwatchers of the Montlake Landfill

In 2003 and 2004, I spent a sequence of Sundays at Seattle's Montlake Landfill—the covered-over landscape described in chapter 5—in an effort to observe and talk with birdwatchers there. I have said a lot in this book about birdwatchers, their politics, and their attitudes; this field work was my attempt to study them in person. In the end, I contacted a dozen birdwatchers at the landfill, and to add to those findings, I circulated an online questionnaire to subscribers of two e-mail discussion lists used by birdwatchers in the Pacific Northwest.[1] With their consent, I recorded the conversations I had with the twelve birdwatchers at the Montlake Landfill and received completed online questionnaires from 225 respondents.

Most of the online respondents to this qualitative study described themselves as fairly experienced birdwatchers (78 percent reported the ability to identify more than one hundred species "off the top of my head"),[2] whereas the birdwatchers I spoke with in person varied in level of experience and reasons for birdwatching. My findings from this qualitative research are in no way representative of all birdwatchers in the Pacific Northwest or elsewhere, but what I discovered helps provide a backdrop for what I have said about birdwatching, field guides, and the environmental imagination.

As a way of concluding *Binocular Vision,* I want to describe what I learned from this group of birdwatchers in and around Seattle in relation to the main themes in this book. Binocular vision, as described here, is rooted in the histories, technologies, and representations of birds in birdwatching field guides. At the same time, my shaping of the term *binocular vision* is meant to describe a much broader approach to encountering and studying a separate, nonhuman "nature." Birds, animals, insects, rocks—these things and others are also sorted and broken down into categories. They are identified, managed, and all too frequently imagined as separate from some of the more pressing environmental crises of our time.

Although there were many correspondences between what I expected to find at the Montlake Landfill and my encounters with the birdwatchers there, there were also some surprises. One I will spend some time describing involves the ways birdwatchers modify their field guides. I take this practice of altering print field guides as metaphorically suggestive of the kinds of dialogues that exist between birdwatching as an off-the-shelf, preconfigured environmental pastime and birdwatching as a potentially more individualized, adaptable, and customized hobby. By discussing the ways many of the birdwatchers I studied inhabit field guides and birdwatching in personalized, eclectic, and varied ways, I suggest that the pastime of birdwatching is in some ways based on altered, variable uptakes of binocular vision that have potential for change.

A major theme in this book has been history. Since the 1880s and '90s, when birdwatching first took shape as a named hobby, what birdwatchers do, why they do it, and the tools birdwatchers use have all changed considerably. These changes have had to do with a range of factors, everything from new print technologies and conventions in field-guide design to developments in environmental conservation and law. When birdwatching first emerged as an environmental pastime, nineteenth-century practitioners did not have competitive birding events to attend, nor did they have portable electronic field guides that played birdsongs at the touch of a button. But they did have narrative guides written by authors who drew on a range of influences to develop sentimental, emotional registers for birdwatching. Birdwatching was first formulated as a complex of feelings intended to make saving birds something that birdwatchers and others would feel strongly about.

One of the birdwatchers I talked with at the Montlake Landfill described her own collection of field guides as a repository for histories about birds and birdwatching. I call her the "Master Birdwatcher" because, on the day I contacted her, she was out giving a birdwatching lesson to a novice birdwatcher:

> *Master Birdwatcher: I even have some of the, uh, um, the old National Geographic bird guides. They're printed in 1938. With colored plates. And those are really fun to look through. They're not exactly field guides, but they're enormous fun because, uh, first of all, the birds all have different names [inaudible] has changed, just reading the text, you can see that things that used to be common aren't, whereas things that maybe were really rare then are better now. It's not all a bad picture either.*

For the Master Birdwatcher, changes in bird names, instituted by the American Ornithologists' Union, are an important feature of birding, and her practice of using historic field guides helps her locate present-day birdwatching in a history of the pastime. In describing a warbler in a nearby tree, she reveals this interest in history:

> *Master Birdwatcher: Isn't that pretty? Yep. Yellow. Now that one's a little bit different. You see, he's got a white throat. There's two kinds of yellow-rumped warblers. Um. One is . . . In fact, they used to be called different species. The one with the white throat is called Mertle Yellow Rumped Warbler, and the other one, with the yellow throat . . . here he is . . . is called an Audubon's, there he goes, and people thought they were a completely different species. Now they think they're the same species, just two variations of color, and, un, they're really really pretty. That was a male.*

In noting that "they used to be called different species," she indicates that bird names and classificatory categories change; keeping track of these changes is part of birdwatching to the Master Birdwatcher.

Another broad theme in this book has been the importance of law to birds and birdwatching, and I discussed law in relation to sentiment because, when it comes to birds, the two have never been entirely separate. Not all environmental laws dealing with birds are dedicated to protecting them, for instance, as bounty laws have enlisted the public in attempts to either exterminate or greatly limit the populations of certain birds (House Sparrows, crows, European Starlings, Bald Eagles, and many others). Similarly, revisions to the Migratory Bird Treaty Act of 1918, in the form of the Migratory Treaty Reform Act of 2004, have made it easier for federal and state agencies to aggressively manage "invasive species" such as the Mute Swan.

Hunting laws enable aggressive management as well. Bird protection, bird management, and birdwatching all rely in some ways on killing birds. On the flip side, of course, is that many avian environmental success stories are associated with U.S. environmental law: the Endangered Species Act of 1973 was so successful in the case of the once-reviled Bald Eagle that the species has since been delisted as an endangered bird. Bald Eagles were once hunted as bounty birds, saved by the mid-twentieth-century environmental movement, and are now part of strong feelings of environmental nationalism.

One birdwatcher I spoke with expressed frustration over what she described as the hype associated with the Bald Eagle. As we spoke, a Bald Eagle alighted in a nearby tree (we were, after all, in the Pacific Northwest):

> *Researcher: Did you get a chance to look at the eagle up above us? [pointing to the Bald Eagle]*
>
> *Birdwatcher: I did. Yeah, uh, I saw the eagle. I, uh, I don't know. I think it's cool to watch them bring their nesting materials over. But—*
>
> *Researcher: Those little sticks and stuff?*
>
> *Birdwatcher: Yeah, I think it's kind of funny, people come out and "Did you see the Eagle? Did you see the Eagle?" I mean, Eagles are cool and everything, but I don't think it's that big of a deal. It's, it's I like their size. I think their size is really kind of impressive. Other than that it's like "Hmm, eh, eagle." I don't know, I guess because I've seen 'em enough, it's not a very big deal.*
>
> *Researcher: It was kind of funny. Just looking at all these people were just circling this tree, looking out for the eagle.*
>
> *Birdwatcher: Yeah, it's insane. Everyone is out here to see the eagle. It's funny. Um. Yeah. And then they're happy. "I guess we'll go now."*
>
> *Researcher: "Our nation's symbol, back to the RV."*
>
> *Birdwatcher: Yeah, exactly. So I, I don't know. That's kind of weird. But, um, I saw it. I wanted to see the nest. It's pretty big but I didn't. I didn't see it. I didn't look more. I just wanted to go get my sandwich.*

As you can see from the exchange, my sentiments were largely in line with this birdwatcher, potentially influencing her responses. That point aside, though, the exchange exemplifies how the birdwatchers I studied incorporate sentiment into their practices.

Although technical field guides may be devoid of such preferences, birdwatchers still map feelings onto birds. Many of the birdwatchers I encountered have developed what I see as their own customized, personalized brands of birdwatching, liking some birds more than others in ways that echo sentiments alive in writing about birds in the nineteenth century. Although this birdwatcher had seen and identified the eagle and although she was interested in the nesting behavior, she was also a bit disenchanted with the Bald-Eagle-fever expressed by the other birdwatchers there. Although all my findings confirm that the technical ways of thinking about and seeing birds are predominant among the birdwatchers I studied, my field research also revealed that birdwatchers still have favorite birds, they still have birds they dislike, and sentiment still plays a role in how they think about and respond to birds.

Another principal theme of this book involves ways of seeing and thinking about birds and nature. By steering several of the book's case studies toward examinations of birds in technologized, polluted, and inhabited landscapes, I argued that being a birdwatcher involves not only interactions with birds but also the development and adoption of an all-encompassing visual, taxonomic disposition toward "nature."

Nature, according to environmentalist field-guide author Jack Griggs, has been thoroughly altered by humans; according to artist Alexis Rockman, industrial society is creating a natural wasteland that is significantly altering life-forms and biodiversity. What is remarkable about Rockman's work is that he questions *seeing* and *looking* as adequate tools for engaging in environmental inquiry. When one watches birds, much that exists in a bird's immediate environment is not seen. In the case of a toxic landfill like the one along the shores of Seattle's Lake Washington, toxicity is invisible. Leachate travels below ground, with birds and birders apparently unharmed up above.

In several of my exchanges with birdwatchers, I heard what I came to define as one-that-got-away narratives. In these narratives (and I identified roughly six of them in my transcripts), birdwatchers construct the sense that the visual approach to studying birds can break down. One birdwatcher, whom I call the New Mexico Birder, describes her one that got away in this way:

> *New Mexico Birder: Like, when I first moved to Santa Fe, I spent the longest time trying to identify what used to be known as the Brown Towhee, which is now the Canyon Towhee. I just couldn't find any pictures that really looked the way it looked.*

And, and . . . but eventually I figured it out. And then, a couple years later, when I got the Sibley guide, I looked in there and I thought "Well, that's what it is!" [laughter]

For this birdwatcher, the one that got away is ultimately identified thanks to what she sees as an improved field guide, but in accounts of other unidentifiable birds, the birdwatchers I spoke with insisted not that a better guide was needed but that the species was simply unidentifiable.

Here is a transcript from one such exchange I had with a birdwatching couple. The man of the couple identified himself as a bird gardener and the woman as someone who likes to go on organized birdwatching trips.

Bird Gardener: We have to kind of attribute it to the fact that the coloration is probably different than what the actual book is, time of year, whatever. We did find one down at Malheur. We used to go to Malheur quite a bit. And, we found one, I think we spent like an hour watching this little brown bird in the middle of the road.

Traveling Birder: It wasn't brown though.

Bird Gardener: [overlapping] No.

Traveling Birder: It was a little bird.

Bird Gardener: [overlapping] It had Elvis Presley sideburns. And it had kind of a bib, kind of like a Flicker, and it had yellow, and it was just a little size of a sparrow, and we've gone through every book we can come across and haven't found it.

Researcher: Bright yellow?

Bird Gardener: Well, it had some bright yellow.

Traveling Birder: It had black and white, right?

Bird Gardener: I remember Elvis Presley sideburns. And it had yellow on its face, as I recall. And I do remember for sure it had kind of like a Flicker [unintelligible]

Researcher: That sounds like a meadowlark to me . . .

Traveling Birder: No, it was smaller than a meadowlark.

Bird Gardener: It wasn't, was it?

Traveling Birder: I think it was black, white, and yellow.

Bird Gardener: Nope. Yep. Probably. It's been a few years.

This one-that-got-away narrative and others like it are important because they capture the birdwatchers as still having a sense of themselves as explorers and discoverers of birds, but also because they show what Michael Lynch and John Law (1999) describe as the limits of Peterson's visual method of bird identification. In these narratives, birds escape the visually oriented birdwatcher; in binocular vision, entire environments and environmental problems escape visual inquiry of the kind bird*watching* endorses.

Since it caught on in the 1890s, birdwatching has been a technologically and textually mediated environmental encounter. Field guides, binoculars, spotting scopes, and a range of other electronic gadgets are currently integral to field identifications. Over the one-hundred-plus years of birdwatching, print field guides and binoculars have been the primary technological artifacts that birdwatchers use, making the primary accomplishment of birdwatching—a positive identification—possible. Today, technologies have changed and multiplied. Many field guides are now digital, available as software for handheld gadgets and online, and birdwatchers have tools capable of making automatic bird identifications.

Several of the birdwatchers I studied in the Pacific Northwest were what might be called Web 2.0 birdwatchers, connecting with one another and sharing information online before and after their birdwatching experiences. Respondents spoke of the information they had gleaned from local Listservs as well as doing searches for various birds using Google. One birdwatcher at the landfill—I call him the Bird Photographer—was particularly technologically savvy, uploading the digital images of birds he took to his website to get identification help from other online viewers. Responding to a question I posed about how many photographs of birds he takes on a single outing, he responded:

> *Bird Photographer: Oh, a couple a hundred maybe. See, what I do is I take the pictures, then I go back, put 'em on the computer, open the book [. . .] That's how I figure 'em out. [. . .] And I'm not very good at doing it in the field anyway. With the quality of the cameras and the lenses, it's wonderful what you can do.*

Using Photoshop, he enhances the images, bringing out some of the color and detail. Birdwatching, as the Bird Photographer shows, is quickly changing. Not only are people able to use new tools to more vividly see and hear birds and to identify them in the field, but technologies and technological networks are redistributing the identification work of birdwatching. In this sense, technobirders are becoming more immediately connected with other like-minded members of their birdwatching communities.

And this community has some interesting forms of variation. The birdwatchers I studied came to birdwatching from a range of positions and backgrounds: some were amateur ornithologists, others environmentalists, and still others amateur historians of their own pastime. One thing that united them all, however, was their dedication to bird study. I cannot emphasize this characteristic enough: the birdwatchers I spoke with and communicated with online all evinced amazingly detailed knowledge about birds.

These birdwatchers are people who know the names of birds, their migration patterns, facts about diet and breeding. Birdwatchers pride themselves on knowing their birds, and what they do not know they want to learn. Birdwatchers also have varied practices: some list the species they see, others travel in search of rare birds, and many alter and customize their field guides. What unites the birdwatchers I studied, though, is this deep interest in all facets of bird life and identification. Birdwatching is a technical, detail-driven pastime, and although I did not get the sense, from the birders I spoke with, that the toxicity of the Montlake Landfill was of grave concern to them or that the indiscriminate hunting of crows in Washington State posed a problem, I did get the sense that bird study was something they were consumed with. Most if not all of the birdwatchers I spoke with found birdwatching immensely rewarding.

MODIFIED FIELD GUIDES

Of the 225 birdwatchers who responded to my online questionnaire, nearly half—105 in all—reported modifying their field guide(s) in some way. Modified field guides are store-bought, mass-produced originals overlaid or reconfigured with personal touches, small additions, or significant alterations. These customized texts are ultimately in excess of the books in their off-the-shelf form, and most modifications indicate how print field guides are a technology that invites customization in ways that the portable electronic and online field guides make difficult. In other words, the books birdwatchers use are very amenable to being written on or otherwise changed.

For the birdwatchers who modify their print field guides, the books in their original form are seen as imperfect, but once the book is improved by the user, the guide both functions more effectively and means more to its owner. Because of the prevalence of modification, to understand one birdwatcher's field guide is not necessarily to understand another's, thus making claims about "field guides" in relationship to individual birdwatchers conditional. Field guides that are written on and physically reconfigured illustrate how birdwatching is defined by purposeful, customized uptakes of the hobby. In what follows, I describe some of the ways that off-the-shelf generic field guides were transformed by my research subjects into personalized technologies, showing how modified guides can become metaphors for the many varieties of birdwatching experiences.

One of the more common ways that birdwatchers modify their print guides is by adding access markers, features of a book that make content more readily and quickly accessible. A birdwatcher I met in the field (I call her Kate) carried a copy of Peterson and Peterson's *A Field Guide to Western Birds* (1990) with her. The book was fastidiously outfitted with numerous translucent purple Post-its, creating a do-it-yourself, impermanent, quick-reference tab system that helped Kate speedily access particular pages and sections of the book. In kind, several of the birdwatchers who responded to my questionnaire reported pasting tabs into their guides.[3]

Respondent 63 puts Post-its "on the pages of birds I am looking for, expecting to see on a trip etc so that I can quickly turn to the page"; Respondent 29 adds "tabs—to quickly find family/genus"; and Respondent 71 says: "I have a few colored tabs that I can move from page to page. If I am comparing two sparrows in the field I move the tabs to those two pages." The addition of tabs may seem like a minor change, but these added access markers indicate the importance many field-guide users place on speed and a personally useful text.

The addition of extra indexes is another common modification, and in recent years aftermarket indexes have become available online. Such indexes can be easily downloaded, printed, and pasted into a field guide. Respondent 86 writes that she or he "glued a general index to the back of

Nat[ional] Geo[graphic field guide] and Sibley to save time. Only need to get families, so index fits on the back. Downloaded and printed these from sites." Several respondents to the online questionnaire described having "downloaded and pasted in a quick index for the big Sibley" (Respondent 20). These custom indexes, available for a range of guides and some through regional Audubon Societies, maximize pathways (via Latin species name, color, or size, for instance) into the often dense content of a guide. Similar to adding tabs, these extra access systems make a guide seem unique through the act of customization, subtly personalizing what it means to use a field guide as a birdwatcher.

Respondent 191 describes an elaborate, one-of-a-kind system of modification in the following response:

> In my National Geographic guide, I've gone through and circled the names next to the pictures of all the birds I've seen, and marked them in the checklist in the back. I've put slashes through the names of all birds (next to their pictures) which don't ever occur in Washington, so that when I'm trying to find out, e.g., what type of sparrow I just saw, I don't waste time looking at pictures of ones that don't occur here (saves time over looking at photo and then range map). Some birds have both a slash and a circle, since I saw them outside of Washington. I make notes next to some pictures as to when and where I saw certain birds. I've put tabs sticking out for the different groups of birds (homemade, not one of those sets you buy—although I've put one of the sets of tabs in a different field guide that I don't use much).

Such a description captures a birdwatcher *enhancing* but not radically altering the guide, making the text quicker on the draw while not disrupting the central impulse of the book to help classify and identify. As print field guides become even more sluggish when compared to highly and multiply-searchable online guides, users are modifying their texts in ways that make them approximate the information architecture of online guides. The modified guides these respondents describe become even more visually and manually scannable texts.

Some modifications are even more dramatic. These changes make using a guide faster, but they also give the book a different feel and character. For instance, several respondents to my online questionnaire described cutting their field guides out of their bindings and having them rebound. Respondent 115 writes: "I cut the binding off of the NGS

[National Geographic Society's *Birds of North America*] and got it spiral bound at Kinko's. That way it stays open to the page I'm working with between birds. I also buy after market indexes. NGS index is not the best." Respondent 19 writes: "NGO [National Geographic's guide]—I had it comb-bound at Kinko's. Easier to handle and use." And Respondent 94 writes: "I had the NG3 [National Geographic, 3rd edition] spiral bound for easier access and to allow me to open it flat. I have also added my own notes related to ID of a species in or near the text account of many species." Slicing a book out of its binding and having it rebound is not something most of us would think to do with the reference materials we use, but these respondents alter the physical mechanics of their guides because the books are tools that they need to have working just right. Perhaps more than any other modification, rebinding reveals this modifying mentality.

Field guides are objects these users care enough about to spend extra time, effort, and money to alter in the ways they describe. This mentality of modification sees the field guide as a book to chop up and punch holes in, write over and recast: raw material used in the making of something better. The guide that is cut from its binding and rebound at a copy shop is a highly instrumental text valued for what it can help users do. This kind of technological modification changes the feel and mechanics of the text as an embodied physical tool, and it is a tweak that shows how much manual mechanics matter to birdwatchers making this hack. Having a guide that "allow[s] me to open it flat" is important to some, demonstrating how subtle maneuvers and daily habits come to matter a great deal. With field guides from the late nineteenth century onward, importance has always been given to cover type, size, feel, and weight; modifying users, though, treat the off-the-shelf book not as a completed product but as a work-in-progress that, in material and technological ways, they coauthor. This customized uptake does not thwart the content of the guide as much as make it more appealing and personalized.

Although adding a spiral binding to a field guide does not exactly change what is said or represented on the printed page of a field guide, some users modify those aspects of their guides as well. Responding to my survey question about modification, Respondent 27 writes: "Mexican and Costa Rican field guides: cut out color plates and bound them separately because the complete books are too cumbersome to use as field guides." Similarly, Respondent 121 writes: "I have taken out the color plates and have them bound separately, to decrease the size of the book." These birdwatchers split their guides in two based on text and image, indicating not only divided ways of reading and thinking about those textual elements, but also the premium many users place on portability. By creating two guides—one image only and the other text only—these birdwatchers render inert how image and text might work together in a relationship. Even so, "decreas[ing] the size of the book" matters to most birdwatchers who carry their field guides with them so much so, in fact, that cutting a book in two makes sense.

Birdwatchers do other little things to make their field guides their own—such as outfitting them with protective, waterproof covers—as part of a family of customizing practices that express mild dissatisfaction with field guides as an off-the-shelf technology. At the same time, there seems to be a satisfaction with the way books of this kind invite modification. Respondent 6 describes having "repaired [a guide] after damage, made the covers water-resistant, put my name in it, updated text after splits [in species classification] and renaming [of species by the American Ornithologists' Union], etc." Respondent 30 reports, "My old Golden Guide fell apart in the late 70's, and I put a ring through it after punching holes in each page."

These changes are personalizing touches—things birdwatchers do to make their mass-produced guides unique and in minor ways representative of their owners. If a field guide can be construed as a symbol for the larger pastime of birdwatching, the modified guide speaks to the ways birdwatching varies—meaning different things to different birdwatchers. There is no one reason birdwatchers do what they do, and this variation and elasticity is a large part of what has made birdwatching such an enduring and widely appealing environmental activity.

I end this book with these examples of modified field guides to emphasize the extent to which bird-identification practices change and vary. Recall that birdwatchers, birders, and twitchers take up birdwatching in slightly different ways, but as this discussion of modification indicates, the variation is even more manifold than that. Members of each one of these birdwatching groups do things in their own, somewhat eclectic, ways. Binocular vision has also varied, changed, and developed since the formation of the hobby in the 1880s and '90s. As my field research has shown, birdwatchers still engage birds with their sentiments, so even though the technical preoccupations of birdwatching predominate, they are not dictatorial.

What is at stake in recognizing birdwatching as an environmental pastime defined by this largely visual disposition I call binocular vision is not just the recognition that birds are seen through a taxonomic, visual lens. In fact, recognizing that binocular vision is a way of seeing birds is a small point. Thinking about binocular vision as manifest in birdwatching helps explain a much larger aspect of our environmental imagination, to return to Lawrence Buell's term.

Binocular vision has become a pervasive way of seeing, compartmentalizing, and conceptualizing the thing many of us still think of as "nature." In binocular vision, the environment is made up of separate objects that are exciting to identify—trees, plants, animals, birds—while being disconnected from one another and most everything else. These individual elements are but representatives to be placed in taxonomic categories, so an individual King Bolete mushroom is not a part of a forest as much as a representative of that type of mushroom. The result is that the larger complex of legal, environmental, and cultural networks that all these nature particles are embedded in and influenced by are attended to much less as one struggles, employing the environmental imagination of binocular vision, to get a grasp on what species something is or is not. Encountering nature with binocular vision means discounting ourselves as part of nature and, as a result, ignoring one of the planet's most significant environmental factors.

The work of some contemporary environmental artists, such as Kim Stringfellow and Chris Jordan, disrupts overly sanitized and singular representations of birds. Stringfellow's project *Greetings from the Salton Sea* (2001–2009) shows the deadly consequences of waterborne pollutants on birds, for instance, and Chris Jordan's photo essay, "Midway: Message from Gyre" (2009), highlights the deadly effect of plastics on birds in the Pacific. Both artists produce what looks very much like wildlife photography, but with a twist: the birds in their images have died as a result of human pollution. Stringfellow and Jordan represent just one option; Jack Griggs and Alexis Rockman some others. Unquestionably, however, new practices for representing birds are clearly needed.

Needed also are changes in birdwatching practices. As birdwatching develops into an increasingly gadget-preoccupied pastime, with birdwatchers establishing vast networks of online connectivity based around data derived in part from citizen science efforts, there is a chance that these networks will sustain not only a taxonomic imagination and binocular vision, but, more robustly, multiple ways of seeing and interacting with birds as connected to the larger environments they are part of. Policy, sentiment, history, and taxonomy can come together in new media encounters with birds. I locate this promise in my discovery that so many birdwatchers modify their field guides.

Birdwatchers already repurpose some of what is presented to them in field guides, customizing their texts to make them faster, yes, but also more personally useful. With textual modification as a core value among so many birdwatchers, I see variation and change in the environmental imagination of birdwatching as likely. One clearly necessary change is the development of a more advocacy-based, broadly aware approach to environmental study and interaction.

NOTES

INTRODUCTION

1. The U.S. Fish and Wildlife Service estimates that 63 million people in the United States feed and/or watch birds (U.S. Fish and Wildlife Service 2000). Some estimates of the number of birdwatchers are as low as 7 million and as high as 70 million. The number 40 million birdwatchers is used by the Cornell Lab of Ornithology (www.birds.cornell.edu/celebration/sponsors).
2. The National Audubon Society was originally established in the 1890s to combat the overhunting of birds (Graham and Buchheister 1992; Orr 1992; Price 1999, 2004).
3. The U.S. Fish and Wildlife Service (2000) estimated that $3.5 billion is spent annually on birdseed, birdhouses, and bird feeders; add to that figure expenditures on field guides, binoculars, spotting scopes, and birdwatching tours, and we see that birdwatching is unquestionably big business.
4. A recent study by twenty-nine scientists, published in the journal *Nature*, argues that nine forms of human-created environmental change threaten to end the current geologic era: climate change, the rate of biodiversity loss (terrestrial and marine), interference with the nitrogen and phosphorus cycles, stratospheric ozone depletion, ocean acidification, global freshwater use, change in land use, chemical pollution, and atmospheric aerosol loading (Rockström et al. 2009). This study complicates the prevailing and much simpler perception of climate change as the only potentially catastrophic environmental problem facing the world today. Climate change is significant, say these scientists, but it is only part of the problem.
5. To develop a historical study of field guides, I have become indebted to several

published cultural histories of birdwatching. Scott Weidensaul's *Of a Feather: A Brief History of American Birding* (2007) is one of the most recent and thorough of these cultural histories, providing a particularly detailed account of the popular *Sibley Guide to Birds* (2000). Stephen Moss's *A Bird in the Bush: A Social History of Birdwatching* (2004) is equally comprehensive in dealing with popular forms of birdwatching as a kind of craze in Great Britain in particular. Jonathan Rosen's *The Life of the Skies: Birding at the End of Nature* (2008) describes many connections between birdwatching and literature. There is a tendency in some histories, however, to develop a masculinist narrative that focuses more on male birdwatchers, ornithologists, and field-guide authors than on females, even though women have contributed enormously to field-guide literature and birdwatching. Some scholars, however, have discussed the important role of women in birdwatching and bird protection. Felton Gibbons and Deborah Strom's *Neighbors to the Birds: A History of Birdwatching in America* (1988) and Deborah Strom's *Birdwatching with American Women: A Selection of Nature Writings* (1986) are two such histories, paying close attention to such important birdwatchers as Rosalie Edge, who founded Hawk Mountain Sanctuary, and Mabel Osgood Wright, the founder of Birdcraft Sanctuary. Daniel Philippon's introduction to his reissued edition of Wright's *The Friendship of Nature: A New England Chronicle of Birds and Flowers* (originally published in 1894) is another such work, making the point that "modern conservationists might recoil in horror at the style of wildlife management practiced at Birdcraft in these early years" (1999, 18). Vera Norwood argues that gender is important because "masculine and feminine" performances in nature writing "help structure different versions of nature" (1999, 53). Equally compelling is Jennifer Price's work on gender, fashion, environmentalism, and the late-nineteenth-century plume trade in *Flight Maps: Adventures with Nature in Modern America* (1999) and "Hats off to Audubon" (2004). Price points not only to the important role that many women had in the creation of the early Audubon movement and bird conservation, but also to how those contributions were en-meshed with gendered politics.

6. Mark Barrow, in *A Passion for Birds: American Ornithology after Audubon* (1998), describes how ornithology took shape as a scientific discipline in North America much later than it did in Europe (as described by Farber 1982/1997).
7. Lynch and Law's 1999 "Pictures, Texts, and Objects" builds directly from their earlier publication: "Lists, Field Guides, and the Descriptive Organization of Seeing: Birdwatching as an Exemplary Observational Activity" (1988).

CHAPTER ONE.
FIELD GUIDES AND THE NEW HOBBY OF BIRDWATCHING

1. Other somewhat earlier examples of natural history writing about birds include Emily Taylor's *Conversations with the Birds* (1850) and C. W. Webber's *Wild Scenes and Song-Birds* (1855).

2. By selecting these two early guides out of all the newly available field guides and texts about birds in the 1880s and 1890s, I do not mean to privilege *Birds through an Opera Glass* and *Birdcraft* over, for instance, Frank Chapman's important and comprehensive guide *Bird Life: A Guide to the Study of Our Common Birds* (1897) or John B. Grant's *Our Common Birds and How to Know Them* (1891). Instead, I mean to analyze two of the more popular guides in this period of genre emergence while focusing rhetorically on two texts that contrast markedly with the quick-reference, taxonomically concentrated guides discussed in the following chapters. In the 1880s and '90s, Florence Merriam and Mabel Osgood Wright were among a subset of the larger group of bird-book authors—the subset including Neltje Blanchan and Olive Thorne Miller—who embedded an emerging ornithological preoccupation with cladistics in more familiar and culturally salient rhetorics of sentiment, emotion, and political emancipation.
3. Korpi also discusses Grant's somewhat later *Our Common Birds and How to Know Them* (1891) as a first American field guide.
4. Henri Lefebvre describes the hobby of photography in a way that aligns with Gelber's reading of hobbies as productive leisure, calling photography "active leisure" and "a cultivated or cultural leisure" (1947/1991, 32). For Lefebvre, leisure is not distinct from remunerative work, but part of a compensatory whole: "Work, leisure, family life and private life make up a whole which we can call a 'global structure' or 'totality' on condition that we emphasize its historical, shifting, transitory nature. . . . Our particular concern will be to extract what is living, new, positive—the worthwhile needs and fulfillments—from the negative elements: the alienations" (42).
5. A decade after the publication of her first field guide, *Birds through an Opera Glass*, Florence Merriam married the naturalist Vernon Bailey, changing her name to Florence Merriam Bailey. Because I refer exclusively to the work she published before marrying, during the 1880s and 1890s, I refer to her as Florence Merriam.
6. The first edition of *Birds through an Opera Glass* was published in 1889 as the third volume in the *Riverside Library for Young People* series, alongside such titles as John Fiske's *The War of Independence* and Horace E. Scudder's *George Washington: An Historical Biography* (Chester 1890, publisher's endpapers). Merriam's guide was then reprinted in 1899 by the Chautauqua Press as part of the Chautauqua Literary and Scientific Circle series, a degree-granting "four-year course of required reading" (Book Club, n.d.). The Chautauqua series included such books as Richard T. Ely's *The Strength and Weakness of Socialism* and Carl Schurz's *Abraham Lincoln: The Gettysburg Speech and Other Papers.*
7. My analysis of *Birdcraft* is of the first reprint of the first edition, although the guide changed slightly in later editions, mainly in terms of gaining prefatory materials and losing costly color plates. By the ninth edition (1936), for instance, *Birdcraft* contained no color plates, several new full-page grayscale images of single birds (as opposed to comparative plates), and a "Map of the Bird" image (p. 34) as had become standard in the genre.
8. *Birdcraft* was Louis Agassiz Fuertes's first major commission of bird art, although

his work would quickly become a hallmark of this era of field guides and natural history texts. By the ninth and final edition of *Birdcraft*, the color plates were dropped (surely as a cost-saving measure) and grayscale versions of Fuertes's plates appeared only sporadically in the text.

9. Several children's books about birds, published in that era, strongly relied on anthropomorphism as well. See Mabel Osgood Wright, Elliot Coues, and Louis Agassiz Fuertes's *Citizen Bird* (1897) and Thornton Burgess and Louis Agassiz Fuertes's *The Burgess Bird Book for Children* (1919).
10. My use of the term *taste* comes from Pierre Bourdieu's use of the term in *Distinction: A Social Critique of the Judgment of Taste* (1984), a sociological study of consumption and taste in modern France. For Bourdieu, taste functions to pit social classes against one another, establishing social economies of high and low based on what different groups dislike and prefer.
11. F. Shuyler's Mathews's *Field Book of Wild Birds and Their Music* (1904) is perhaps the most dramatic piece of evidence of this domestic passion for and interest in birdsongs, a field guide dedicated to rendering bird "music" on the musical scale. The rise in the quality of optics and fall in the price of binoculars seem at least partly responsible for the diminished importance given to birdsongs in field guides and birdwatching circles by the middle of the twentieth century.
12. Wright tries to improve bird reputations most notably in relation to water birds. Recall that ducks, geese, gulls, and other shore birds were completely left out of Merriam's field guide. Wright writes: "When you think of the Water-birds, you say, perhaps, that they are uninteresting, have no song, and inhabit marshy and desolate places. . . . [They are] only game birds and so much food. . . . This is because you have regarded them as mere merchandise, and have never seen or considered them as living birds" (1895, 21).
13. This theme of neighborliness is notable, too, in Wright's book of natural history, *The Friendship of Nature: A New England Chronicle of Birds and Flowers* (1894).

CHAPTER TWO.
NUISANCE BIRDS, FIELD GUIDES, AND ENVIRONMENTAL MANAGEMENT

1. In this chapter, I describe a range of avian environmental management practices. Although the term *management* refers to both the support of lagging species and the control of superabundant ones, the term is most often used to refer to control.
2. The list of birds managed as nuisance birds is expansive and currently includes such species as Monk Parakeets because they build large nests that sometimes harm power lines, woodpeckers of several kinds because they drill into wooden structures, and pigeons largely because of their excrement. A nearly complete list of nuisance birds is given in Witmer et al. "Management of Invasive Vertebrates in the United States: An Overview" (2007).
3. Frank Getz Ashbrook published several small field guides in the early 1930s that

were indicative of the trend toward more compact, technical guides (Ashbrook 1931).

4. Chapman's guides include *Handbook of Birds of Eastern North America* (1895) and *Bird-Life: A Guide to the Study of Our Common Birds* (1897).
5. "Birds of prey" was a common catch-all term used to refer to hawks, eagles, and some owls. Other birds such as Loggerhead Shrikes are certainly predators but are generally not referred to with the term "birds of prey."
6. In 2007, the Bald Eagle population recovered to the extent that the species was removed from listing under the Endangered Species Act. Bald Eagles remain protected, however, under what has been renamed the Bald and Golden Eagle Protection Act. Amendments to the latter piece of legislation now allow for limited hunting of both the Bald and Golden Eagles, and in some cases the birds can be taken to be used in falconry (U.S. Fish and Wildlife Service 2008).
7. The introduction of species to North America did not happen on an informal basis, but rather was the result of organized efforts. The American Acclimatization Society is the most well known group that, in the nineteenth century, was dedicated to introducing European plants and animals to North America. The society introduced several species of birds (for example, the Sky Lark, European Starling, House Sparrow, and Mute Swan) into North America. As John Leland describes it, the importation of large numbers of birds by such societies (three thousand individual birds in Ohio alone between 1873 and 1874) was considered "cutting edge science" at the time, aimed at making the landscape more beautiful while attempting to use certain birds to fight pests (Leland 2005, 165).
8. Egg addling generally involves one or more of the following: coating the eggs in a thin film of oil, shaking them, or poking a small hole in the eggshell.
9. Avicide of this kind is generally justified by the management philosophy known as *compensatory mortality*. Discussions of compensatory mortality can be found in Weller (1988, 576) and Mech and Boitani (2003, 156–57).
10. Bounties for pest birds do remain on the books in Michigan and North Dakota, for example.
11. Although the Tamaulipas Crow was once found in south Texas, this species has now largely disappeared within the forty-eight contiguous states. The species is still highly sought after by birdwatchers, though, many of whom visit the Brownsville, Texas, dump in search of what was previously called the "Mexican Crow."

CHAPTER THREE.
PICTURING BIRDS IN ALTERED LANDSCAPES

1. The illustrations of birds in Jack Griggs's *All the Birds of North America* were created by a group of twelve illustrators.
2. John Perlock's print illustrations for GE's ecomagination series won a bronze World Press Award in 2007.

3. In many ways, GE has an egregious environmental track record. To cite just a few examples, the corporation was found responsible for dumping PCBs into the Hudson River where, as a result, a large cleanup operation is under way by the U.S. Environmental Protection Agency (Environmental Protection Agency 2008). At the Hanford Nuclear Reservation in Washington State, GE conducted research that involved intentionally sending radioactive pollutants into the air; one radioactive cloud traveled as far as California (Multinational Monitor 2001; Cray 2006).
4. Although Jennifer Baker has described GE's use of Audubon's work as a kind of reference to the way Audubon "anticipated the modern conservation movement" (2006, para. 12), most scholars have not characterized Audubon as a proto-environmentalist. Audubon not only participated in indiscriminate "industrial-style killings" of birds (Irmscher 1999, 214) in the process of collecting specimens for his *Birds of America* (completed in 1838), but killed some of his specimens slowly over the course of several days (Irmscher 1999, 206–35). Audubon has been portrayed as more of a capitalist than an environmentalist, with one biographer noting that Audubon became a bird illustrator only after failing at business. Bird illustration for Audubon was a kind of second business, and he took great care to price his work exorbitantly, selling complete sets of prints for $1,000 apiece (Welker 1955, 60–63; Irmscher 1999, 198–99).
5. In another GE ad, this one an interactive online banner ad, GE repeats the specious claim that its engines are friendly to birds. This banner ad describes GE's GEnx jet engine as able to "share the air" with birds (General Electric Corporation 2005a).
6. A 355-page report commissioned by the Federal Aviation Administration and the U.S. Department of Agriculture is by far the most comprehensive published treatment of the aviation problem called bird strike, which is the name used to designate when any part of an aircraft hits a bird. The authors of *Wildlife Hazard Management at Airports: A Manual for Airport Personnel* (second edition, published in 2005), Edward Cleary and Richard Dolbeer, describe the seriousness and scope of bird strike, paying particular attention to methods for mitigating the problem.
7. Examples of cartoons featuring birdwatchers in the *New Yorker* can be found at the magazine's cartoon archive, www.cartoonbank.com; search "bird watching."
8. In Michael Gitlin's film *The Birdpeople* (2004)—a documentary about birdwatchers, ornithologists, and bird banders—this kind of reversal of the birdwatching gaze is similarly explored. Gitlin investigates these three groups of people who are obsessed with birds, and to do so he not only interviews them, but *watches* them. Throughout the film, Gitlin's camera rests on the subjects of the documentary, quietly staring at them in a way that mimics how people interested in birds study their subjects. Jaimie Baron, writing about the film, has described the documentary as "purporting to be about people who watch, band, stuff, and search for birds [while] *The Birdpeople*, in fact, turns the birdpeople's own archival strategies

back on themselves. For one thing, the film is as much about watching people as it is about watching birds" (2007, 20).

CHAPTER FOUR.
TECHNOJUMPING INTO ELECTRONIC FIELD GUIDES

1. I use the term *vocalization* instead of simply *birdsongs* to emphasize that birds produce both songs and calls and that both sets of sounds can function as species markers.
2. I justify the use of the term *sham object* because none of these portable electronic devices identifies all American birds (unlike the print counterparts they mean to displace), and much of the seemingly new information in portable electronic guides is merely derived from print guides.
3. Some bird lovers perusing the eNature.com guide might be shocked to learn that Birdbarrier.com specializes in both "Bird-Flite Spikes" and "Bird-Shock Flex-Track," two painful if nonlethal ways to keep Rock Doves from perching on or around private property.
4. The Biodiversity Institute's *Internet Field Guide to Birds* is another good example.

CHAPTER FIVE.
BIRDING ON TOXIC LAND

1. The Migratory Bird Treaty Act came out of an agreement between Great Britain (acting for Canada) and the United States in 1918. The act has since been multiply amended, with additional countries signing on. The basic intention of the act is to protect migrating birds, although provisions in the act allow for regulated hunting. The U.S. Fish and Wildlife Service has detailed information about the act online at www.fws.gov/laws/lawsdigest/migtrea.html. The Endangered Species Act was passed some fifty years later, covers plants and animals, and allows for the protection of entire ecosystems. For more information on this act, see the U.S. Fish and Wildlife Service's website at www.fws.gov/laws/lawsdigest/ESACT.HTML.
2. Examples of such lists can be found by searching the social networking photo site Flickr. See flickr.com/search/?q=bird+list for examples.
3. Authors such as William Cronon, Anne Whiston Spirn, and Kenneth R. Olwig—all contributors to the collection *Uncommon Ground* (Cronon 1995)—have discussed how even the most "natural," "untouched" landscapes are often highly managed, planned, and even designed spaces.
4. Robert Markley argues that the concept of sustainability is both unfounded and, as a concept, "masks sets of jostling presuppositions about what systems are and how they function" (2007, 7).
5. For example, birding relies on a range of consumer goods, although it is certainly

no worse than most other forms of environmental sporting. Notably, the field guides that birders use are made with wood pulp, glues, petrochemical-based inks, and kaolinite (the clay used to make coated papers). Kaolinite, for instance, is extracted from open-pit mines internationally. The high-grade optics that birders use require mined minerals, rubbers, and plastics made from processed petrochemicals. Until the creation of glass such as Nikon's Eco-glass, the glass used in binoculars and spotting scopes was manufactured with the heavy metals arsenic, lead, and cadmium. As of 2007, however, Nikon reported that 97 percent of the glass it produces is free of these pollutants (Nikon Corporation 2007).

6. The rediscovery of the Ivory-billed Woodpecker (*Campephilus principalis*) has been hotly debated since 2005, with all evidence of the species' existence being scrutinized by scientists and amateur birders alike. For an overview of this debate, see David Allen Sibley's "Ivory-billed Woodpecker—Status Review" online at sibleyguides.blogspot.com/2007/10/ivory-billed-woodpecker-status-review.html.
7. The American Bird Conservancy, National Geographic Society, and National Audubon Society all have field guides, with the National Audubon Society promoting multiple guides.
8. Although most participants in the World Series of Birding travel the state by car, more sustainable forms of transportation are being encouraged by the event's new Carbon Footprint Cup (www.birdcapemay.org/wsob.shtml).
9. International Migratory Bird Day (IMBD) is a newly distinguished "day," established in 1993 by the Cornell Lab of Ornithology. Although called a "day," IMBD does not necessarily fall on any one day but is instead linkable, through promotion materials, to bird and birding events throughout the year. More on IMBD can be found at birdday.org, a website sponsored by the spin-off nonprofit organization that manages IMBD, Environment for the Americas. IMBD should not be confused with other days aimed at raising awareness about birds. Bird Day, for instance, was established to raise awareness about birds and connected, in 1894, to Arbor Day. Celebration of Bird Day seems largely to have fallen into desuetude. There is another Bird Day, which falls in January and focuses on birds in captivity, as well as World Migratory Bird Day, described at worldmigratorybirdday.org.
10. The federal Superfund program is the result of the 1980 Comprehensive Environmental Response, Compensation, and Liability Act and is under the auspices of the EPA. A history of Superfund, a federally funded program aimed at cleaning up the nation's worst toxic sites, can be found at the EPA's website: www.epa.gov/superfund/20years/preface.htm. The EPA maintains a searchable data-base including reports on individual Superfund sites at www.epa.gov/superfund.
11. Superfund sites in Sussex County, New Jersey include A. O. Polymer NJD030253355, Accurate Forming NJD047354832, Ames Rubber NJD002389468, Ames Rubber Corp. Wantage Plant NJD982796450, Metaltec/Aerosystems NJD002517472, MKY Wolf Lake Site NJD986627396, and NY Susquehanna and Western RR Property NJ0000104919 (Environmental Protection Agency 2007c).
12. Blake Scott describes a similar tactic in the pharmaceutical industry with the publication of public citizenship reports (Scott 2009).

13. Books about big-year birding include but are not limited to Kenn Kaufman's *Kingbird Highway: The Story of a Natural Obsession That Got a Little Out of Hand* (1997), Pete Dunne's *The Feather Quest: A North American Birder's Year* (1999), and Mark Obmascik's *The Big Year: A Tale of Man, Nature, and a Fowl Obsession* (2004).
14. The adventures of this bicycling "bird-year" family were described on their blog, birdyear.blogspot.com. The family identified 548 individual species in all.
15. Sewage ponds are alternatively referred to as waste treatment facilities, wastewater plants, municipal waste stabilization ponds, or simply lagoons.
16. Zimmerling cites a study that finds that, among nine orders of arthropods (mainly insects) found at sewage ponds, eight were measured to be between 114 percent and 865 percent larger at sewage treatment facilities than in the related habitat of a beaver pond (Zimmerling 2006, 5–6).

CONCLUSION
THE BIRDWATCHERS OF THE MONTLAKE LANDFILL

1. The list TWEETERS is hosted by the University of Washington in Seattle, and OBOL (Oregon Birders Online) is hosted by Oregon State University in Corvallis.
2. Many respondents reported a deep commitment to birdwatching, with 89 percent reporting that they "traveled to find a rare bird or birds." Similarly, 79 percent had participated in organized rare bird counts, 91 percent kept lists of some kind, and 97 percent reported having field guides (several reported owning ten or more guides). A majority of the respondents (66 percent) said that birds were an aspect of their daily lives in more ways than during formal birdwatching outings, and there was a general agreement among respondents that field guides were not the only tool they used to identify birds (friends, experts, the Internet, and other resources were used as well).
3. The fifth edition of the National Geographic Society's *Field Guides to the Birds of North America* (Dunn and Alderfer 2006) includes seven tabs to direct users to hawks, sandpipers, gulls, flycatchers, warblers, sparrows, and finches. In a sense, this field guide incorporates the vernacular practice of modifying a field guide with improvised tabs.

WORKS CITED

Alaska Department of Fish and Game. 1994. Eagles. www.adfg.state.ak.us/pubs/notebook/bird/eagles.php.

Allen, T. B., and C. Hottenstein. 1983. *Field Guide to the Birds of North America*, 1st ed. Washington, D.C.: National Geographic Society.

———. 1987. *Field Guide to the Birds of North America*, 2nd ed. Washington, D.C.: National Geographic Society.

———. 1999. *Field Guide to the Birds of North America*, 3rd ed. Washington, D.C.: National Geographic Society.

———. 2002. *Field Guide to the Birds of North America*, 4th ed. Washington, D.C.: National Geographic Society.

American Birding Association. 2009. ABA code of ethics. www.aba.org/about/ethics.html.

Arnal, Yolanda Texera. 2002. The beginnings of modern ornithology in Venezuela. *The Americas* 58 (4): 601–22.

Arthur's Home Magazine. 1854. August 4 (2): 88–89; September 4 (3): 168–71; December 4 (6): 440–42.

Ashbrook, Frank G. 1931. *The Red Book of Birds of America*, vol. 1. Racine, Wisc.: Whitman Publishing.

Atkinson, Michael. 2009. Parkour, anarcho-environmentalism, and poesis. *Journal of Sport and Social Issues* 33 (2): 169–95.

Atlantic Flyway Council. 2003. Atlantic flyway Mute Swan management plan. Snow Goose, Brant, and Swan Committee, Atlantic Flyway Technical Section. www.pgc.state.pa.us/pgc/lib/pgc/swans/pdf/mute_swan_plan.pdf.

Audubon, John James. 1840. *The Birds of America, from Drawings Made in the United States and Their Territories*. Philadelphia: J. B. Chevalier.

———. 1999. *John James Audubon: Writings and Drawings*. Edited by Christoph Irmscher. New York: Literary Classics of the United States.

Baily, William L. 1869. *Our Own Birds: A Familiar Natural History of the Birds of the United States*. Edited by Edward D. Cope. Philadelphia: J. B. Lippincott.

Baird, Spencer F., Thomas M. Brewer, and Robert Ridgway. 1874. *History of North American Birds*. Boston: Little, Brown.

Baker, Jennifer. 2006. The making of John James Audubon. *William and Mary Quarterly* 63 (1), www.historycooperative.org/journals/wm/63.1/br_2.html.

Banks, Richard C., R. Terry Chesser, Carla Cicero, Jon L. Dunn, Andrew W. Kratter, Irby J. Lovette, Pamela C. Rasmussen, J. V. Remsen Jr., James D. Rising, and Douglas E. Stotz. 2007. Forty-eighth supplement to the American Ornithologists' Union check-list of North American birds. *The Auk* 124 (3), www.aou.org/checklist/suppl/AOU_checklist_suppl_48.pdf.

Bargmann, Dorie. 2005. Thirteen more ways of looking at a blackbird. *Prairie Schooner* 79 (3): 136–43.

Barnes, Barbara. 2009. "Everybody wants to pioneer something out here": Landscape, adventure, and biopolitics in the American Southwest. *Journal of Sport and Social Issues* 33 (3): 230–56.

Barnes, Simon. 2005. *How to Be a Bad Birdwatcher*. New York: Pantheon.

Baron, Jaimie. 2007. Contemporary documentary film and "archive fever": History, the fragment, and the joke. *The Velvet Light Trap* 60 (Fall): 13–24.

Barrow, Mark V. 1998. *A Passion for Birds: American Ornithology after Audubon*. Princeton: Princeton University Press.

———. 2002. Science, sentiment, and the specter of extinction: Reconsidering birds of prey during America's interwar years. *Environmental History* 7 (1): 69–98.

Battalio, John T. 1998. *The Rhetoric of Science in the Evolution of American Ornithological Discourse*. Stamford, Conn.: Ablex.

Baudrillard, Jean. 1994. *Simulacra and Simulation*. Translated by Sheila Faria Glaser. Ann Arbor: University of Michigan Press.

Bazerman, Charles. 1999. *The Languages of Edison's Light*. Cambridge: MIT Press.

Bean, Michael J., and Melanie J. Rowland. 1997. *The Evolution of National Wildlife Law*. Westport, Conn.: Praeger.

Beans, Bruce E. 1997. *Eagle's Plume: The Struggle to Preserve the Life and Haunts of America's Bald Eagle*. Lincoln: University of Nebraska Press.

Beletsky, Les, and Jon L. Dunn. 2006. *Bird Songs: 250 North American Birds in Song*. San Francisco: Chronicle Books.

Bernzweig, Carl. 2003. World Series of Birding 2003 forum. Message "Where'd you get your Kingfisher?" *New Jersey Audubon*. www.njaudubon.org/Tools.Net/Forum/TopicDetail.aspx?ftk=212.

Binns, Adrian. 2003. Delaware Valley Ornithological Club report: World Series of Birding. www.dvoc.org/WSB/WSB2003.htm.

Birdsong Identiflyer. 2007. Testimonials. www.identiflyer.com.

Black-footed Albatross. enature.com/fieldguides/detail.asp?recnum=BD0001.

Blanchan, Neltje. 1897. *Bird Neighbors: An Introductory Acquaintance with One Hundred and Fifty Birds Commonly Found in the Gardens, Meadows, and Woods about Our Homes.* Garden City, N.Y.: Doubleday.

Blum, Ann Shelby. 1993. *Picturing Nature: American Nineteenth-Century Zoological Illustration.* Princeton: Princeton University Press.

Book Club (Chautauqua literary and scientific circle). No date. Chautauqua Institute. www.chautauqua-inst.org/clsc.html.

Boothroyd, Malkolm, Wendy Boothroyd, and Ken Boothroyd. 2007–2008. Bird year, blog. birdyear.blogspot.com.

Borden, Iain. 2001. *Skateboarding, Space, and the City: Architecture and the Body.* Oxford: Berg.

Born Free. 2008. State Wildlife Bounty Laws by State. *Born Free USA and Animal Protection Institute.* www.api4animals.org/b4a2_bounty.php.

Bourdieu, Pierre. 1984. *Distinction: A Social Critique of the Judgment of Taste.* Cambridge: Harvard University Press.

Brand, Albert. 1934. *Songs of Wild Birds* (vinyl record). New York: Thomas Nelson and Sons.

Brooks, Kevin, Cindy Nichols, and Sybil Priebe. 2004. Remediation, genre, and motivation: Key concepts for teaching with weblogs. In *Into the Blogosphere: Rhetoric, Community, and Culture of weblogs,* ed. Laura Gurak, Smiljana Antonijevic, Laurie Johnson, Clancy Ratliff, and Jessica Reyman. blog.lib.umn.edu/blogosphere/remediation_genre.html.

Buell, Lawrence. 1995. *The Environmental Imagination: Thoreau, Nature Writing, and the Formation of American Culture.* Cambridge: Belknap Press of Harvard University Press.

Burgess, Thornton, and Louis Agassiz Fuertes. 1919. *The Burgess Bird Book for Children.* Boston: Little, Brown.

Burroughs, John. 1871. *Wake Robin.* New York: Hurd and Houghton.

Carson, Rachel. 1962. *Silent Spring.* Boston: Houghton Mifflin, Riverside Press.

Catesby, Mark. 1731. The *Natural History of Carolina, Florida and the Bahama Islands: Containing the FIGURES of BIRDS, BEASTS, FISHES, SERPENTS, INSECTS, and PLANTS: Particularly, the FOREST-TREES, SHRUBS, and Other PLANTS, Not Hitherto described, Or Very Incorrectly Figured by the Authors. Together with Their DESCRIPTIONS in English and French. To Which, Are Added OBSERVATIONS on the AIR, SOIL and WATERS: With Remarks upon AGRICULTURE, GRAIN, PULSE, ROOTS, &c. TO the whole, Is Prefixed a New and Correct Map of the Countries Treated of.* London: By the Author.

Catesby, Mark, and Alan Feduccia. 1985. *Catesby's Birds of Colonial America.* Chapel Hill: University of North Carolina Press.

Center for Urban Horticulture. No date. History of the Union Bay Natural Area. depts.washington.edu/ubna/history.htm.

Chapman, Frank M. 1895. *Handbook of Birds of Eastern North America, with Keys to the Species, and Descriptions of Their Plumages, Nests, and Eggs, Their Distribution and Migrations.* New York: D. Appleton.

———. 1897. *Bird-Life: A Guide to the Study of Our Common Birds.* New York: D. Appleton.

Chester, E. 1890. Girls and women—publisher's end papers listing other volumes in the series. New York: Houghton Mifflin.

Chu, Miyoko. 2009. World Series of Birding: Sneak preview from New Jersey. Round Robin, the Cornell Blog of Ornithology. birdsredesign.wordpress.com/2009/05/07/world-series-of-birding-sneak-preview-from-new-jersey.

Cleary, Edward C., and Richard A. Dolbeer. 2005. Wildlife hazard management at airports: A manual for airport personnel, 2nd ed. Federal Aviation Admin-istration and United States Department of Agriculture. digitalcommons.unl.edu/icwdm_usdanwrc/133/.

Clement, Roland. 2003. Comments on Fitzpatrick. *The Auk* 120 (3): 915.

Coakley, Jay. 2006. The sociology of sport: What is it and why study it? In *Sports in Society: Issues and Controversies,* 2–28. New York: McGraw-Hill.

Coe, Richard, Lorelei Lingard, and Tatiana Teslenko, eds. 2002. *The Rhetoric and Ideology of Genre: Strategies for Stability and Change.* Cresskill, N.J.: Hampton Press.

Congressional Budget Office. 2004. S. 2547: Migratory Bird Treaty Reform Act of 2004. www.cbo.gov/doc.cfm?index=5606&type=0.

Cordell, H. Ken, and Nancy G. Herbert. 2002. The popularity of birding is still growing. *Birding* (February): 54–61.

Cornell Lab of Ornithology. 2002. How do you tell a Fish Crow *Corvus ossifragus* from an American Crow *Corvus brachyrhynchos*? www.birds.cornell.edu/crows/FishCrow.htm.

———. 2003. Mute Swan. *All about Birds Bird Guide.* www.birds.cornell.edu/AllAboutBirds/BirdGuide/Mute_Swan_dtl.html.

———. 2007. World Series of Birding big-day highlights. www.birds.cornell.edu/wsb/2007-highlights.

Coues, Elliot. 1872. *Key to North American Birds.* New York: Dodd and Mead; Boston: Estes and Lauriat.

Cray, Charlie. 2006. General Electric. *CorpWatch: Holding Corporations Accountable.* www.corpwatch.org/section.php?id=16.

Cronon, William., ed. 1995. *Uncommon Ground: Toward Reinventing Nature.* New York: W. W. Norton.

Curtin, Jeremiah. 1923. *Seneca Indian Myths.* Charleston, S.C.: Forgotten Books.

Daily Show, The. 2000. "The World Series of Birding." Video. July 18. www.thedailyshow.com/watch/tue-july-18-2000/the-world-series-of-birding.

Daston, Lorraine, and Peter Galison. 1992. The image of objectivity. *Representations* 40: 81–128.

DeLuca, Kevin M. 1999. *Image Politics: The New Rhetoric of Environmental Activism.* New York: Guilford Press.

DeLuca, Kevin M., and Jennifer Peeples. 2002. From public sphere to public screen: Democracy, activism, and the "violence" of Seattle. *Critical Studies in Media Communication* 19 (2): 125–51.

Dolbeer, Richard A., and Sandra E. Wright. 2008. Wildlife strikes to civil aircraft in the United States 1990–2007. *FAA National Wildlife Strike Database*, no. 14. wildlife-mitigation.tc.faa.gov/public_html/index.html#access.

Dorsey, Kurkpatrick. 1998. *The Dawn of Conservation Diplomacy: U.S.–Canadian Wildlife Protection Treaties in the Progressive Era.* Seattle: University of Washington Press.

Duncan, Neil. 1978. The effects of culling Herring Gulls (*Larus argentatus*) on recruitment and population dynamics. *Journal of Applied Ecology* 15 (3): 697–713.

Dunn, Jon L., and Jonathan Alderfer, eds. 2006. *Field Guide to the Birds of North America*, 5th ed. Washington, D.C.: National Geographic Society.

Dunn, Katherine. 1996. Eden rocks: The art of Alexis Rockman. *ArtForum* 34 (6): 72–75, 109.

Dunne, Pete. 1999. *The Feather Quest: A North American Birder's Year.* New York: Mariner Books.

Elman, Robert. 1974. *The Hunter's Field Guide to the Game Birds and Animals of North America.* New York: Knopf.

enature.com. No date. www.enature.com.

Engeström, Yrjö. 2006. Development, movement, and agency: Breaking away into mycorrhizae activities. *CHAT Technical Reports* 1: 1–44. www.chat.kansai-u.ac.jp/publications/tr/v1_1.pdf.

Environmental News Service. 2003. Criminal claims over Florida bird fatalities resolved. www.ens-newswire.com/ens/oct2003/2003-10-08-04.asp.

Environmental Protection Agency. 1972. DDT ban takes effect. EPA press release. www.epa.gov/history/topics/ddt/01.htm.

———. 1975. DDT regulatory history: A brief survey (to 1975). *EPA Report.* www.epa.gov/history/topics/ddt/02.htm.

———. 2007a. New Jersey sites by name. www.epa.gov/region02/cleanup/sites/njtoc_name.htm.

———. 2007b. Roebling Steel Co. EPA fact sheet. www.epa.gov/superfund/sites/nplfs/fs0200439.pdf.

———. 2007c. Superfund site information. www.epa.gov/superfund/sites/cursites.

———. 2008. Hudson River PCBs. www.epa.gov/hudson.

Farber, Paul Lawrence. 1982/1997. *Discovering Birds: The Emergence of Ornithology as a Scientific Discipline, 1760–1850.* Baltimore: Johns Hopkins University Press.

Fine, Gary Alan, and Lazaros Christoforides. 1991. Dirty birds, filthy immigrants, and the English Sparrow war: Metaphorical linkage in constructing social problems. *Symbolic Interaction* 14 (4): 375–93.

Fisher, Philip. 1985. *Hard Facts: Setting and Form in the American Novel.* New York: Oxford University Press.

Fitzpatrick, John W. 2005. The Ivory-billed Woodpecker still lives. *Living Bird* 24: 3.www.birds.cornell.edu/Publications/LivingBird/Summer2005/still_lives.html.

Forbes, Linda C., and John M. Jermier. 2002. The institutionalization of bird protection: Mabel Osgood Wright and the early Audubon movement. *Organization and Environment* 15 (4): 458–65.

Franklin, Benjamin. 1784. Letter to Mrs. Sarah Bache. www.webexhibits.org/daylightsaving/franklin2.html.

Freedman, Aviva, and Peter Medway, eds. 1994. *Genre and the New Rhetoric.* London: Taylor and Francis.

Gamber, Wendy. 1997. *The Female Economy: The Millinery and Dressmaking Trades, 1860–1930.* Champaign: University of Illinois Press.

Gelber, Steven M. 1999. *Hobbies: Leisure and the Culture of Work in America.* New York: Columbia University Press.

General Electric Corporation. 2005a. Advertisement for GEnx jet engine. www.ge.com/files/usa/en/company/companyinfo/advertising/bird_300x250_30k_15s_p1.html.

———. 2005b. Driving GE ecomagination with the low-emission GEnx jet engine. Press release. www.geae.com/aboutgeae/presscenter/genx/genx_20050720.html.

Gibbons, Felton, and Deborah Strom. 1988. *Neighbors to the Birds: A History of Birdwatching in America.* New York: W. W. Norton.

Gitlin, Michael, producer and director. 2004. *The Birdpeople.* 61-minute film. Distributed by Flat Surface Films.

Gleason, William A. (1999). *The Leisure Ethic: Work and Play in American Literature, 1840–1940.* Stanford: Stanford University Press.

Gough, G. A., J. R. Sauer, and M. Iliff. 1998. Patuxent bird identification info center. Version 91.7. www.mbr-pwrc.usgs.gov/Infocenter.

Graham, Frank, and Carl Buchheister. 1992. *The Audubon Ark: A History of the National Audubon Society.* Austin: University of Texas Press.

Grant, John Beveridge. 1891. *Our Common Birds and How to Know Them.* New York: C. Scribner's Sons.

Gregg, Ian. 2005. Mute Swan history. Pennsylvania Game Commission. www.portal.state.pa.us/portal/server.pt?open=514&objID=623106&mode=2.

Griggs, Jack L. 1997. *All the Birds of North America: American Bird Conservancy's Field Guide,* 1st ed. New York: Harper Perennial.

Gross, Alan G. 1990. *The Rhetoric of Science.* Cambridge: Harvard University Press.

Guillemette, Magella, and Pierre Brousseau. 2001. Does culling predatory gulls enhance the productivity of breeding common terns? *Journal of Applied Ecology* 38: 1–8.

Hanks, Reuel. 2000. A separate space? Karakalpak nationalism and devolution in post-Soviet Uzbekistan. *Europe-Asia Studies* 52 (5): 939–53.

Harrison, E. Z., S. R. Oakes, M. Hysell, and A. Hay. 2006. Organic chemicals in sewage sludges. *Science of the Total Environment* 367: 481–97.

Hartnett, Stephen. 2002. Fanny Fern's 1855 *Ruth Hall*, the cheerful brutality of capitalism, and the irony of sentimental rhetoric. *Quarterly Journal of Speech* 88 (1): 1–18.

Heise, Ursula K. 2006. The hitchhiker's guide to ecocriticism. *PMLA* 121: 503–16.

———. 2008. Ecocriticism and the transnational turn in American studies. *American Literary History* 20 (1–2): 381–404.

Humane Society. 2008. Frequently asked questions about Maryland's Mute Swan management policy. www.hsus.org/wildlife/issues_facing_wildlife/faq_md_swan.html.

Hurrell, Andrew. 1993. Brazil and the international politics of Amazonian deforestation. In *Resource Management in Developing Countries*, ed. Bleshawar Thakur. Delhi: Concept Publishing.

Indiana University. 2006. Studies find heart deformities, higher mortality rates in PCB-affected wildlife. IU Newsroom. newsinfo.iu.edu/news/page/normal/3224.html.

Irmscher, Christoph. 1999. *The Poetics of Natural History*. New Brusnwick, N.J.: Rutgers University Press.

Jewitt, Carey, and Rumiko Oyama. 2001. Visual meaning: A social semiotic approach. In *Handbook of Visual Analysis*, ed. Carrey Jewitt and Theo Van Leeuwen, 134–56. Thousand Oaks, Calif.: Sage.

Jewitt, Carey, and Theo Van Leeuwen. 2001. *Handbook of Visual Analysis*. Thousand Oaks, Calif.: Sage.

Jordan, Chris. 2009. Midway: Message from Gyre. *Chris Jordan Photographic Arts*. www.chrisjordan.com.

Karnicky, Jeffrey. 2004. What is the Red Knot worth? Valuing human/avian interaction. *Society and Animals Journal of Human-Animal Studies*, 12 (3). www.psyeta.org/sa/sa12.3/karnicky.shtml.

———. 2007. "Included in this classification": Encoding American birds. Paper presented at the annual meeting of the Society for Literature, Science, and the Arts. www.slsa07.com/proposals.html.

Kaufman, Kenn. 1997. *Kingbird Highway: The Story of a Natural Obsession That Got a Little Out of Hand*. Boston: Houghton Mifflin.

———. 2000. *Birds of North America*. New York: Houghton Mifflin.

Kofalk, Harriet. 1989. *No Woman Tenderfoot: Florence Merriam Bailey, Pioneer Naturalist*, 1st ed. College Station: Texas A&M University Press.

Korpi, Raymond. 1999. *"A Most Engaging Game": The Evolution of Bird Field Guides and Their Effects on Environmentalism, Ornithology, and Birding, 1830–1998*. Ph.D. diss., Washington State University, American Studies Program.

Kress, Stephen W. 1983. The use of decoys, sound recordings, and gull control for re-establishing a tern colony in Maine. *Colonial Waterbirds* 6: 185–96.

LaDeau, Shannon, A. Marm Kilpatrick, and Peter P. Marra. 2007. West Nile virus emergence and large-scale declines of North American bird populations. *Nature* 447 (7): 710–14.

Laubin, Reginald, and Gladys Laubin. 1989. *Indian Dances of North America: Their Importance to Indian Life.* Norman: University of Oklahoma Press.

Lefebvre, Henri. 1947/1991. *Critique of Everyday Life*, vol. 1. London: Verso.

Leland, John. 2005. *Aliens in the Backyard: Plant and Animal Imports into America.* Columbia: University of South Carolina Press.

Lubbers, Klaus. 2000. Constructing American national identity: The case of the Bald-Headed Eagle in early folk art. In *Negotiations of America's National Identity*, vol. 2, ed. Roland Gagenbüchle and Josef Raab, 8–49. Tübingen, Germany: Stauffenburg Verlag Brigitte Narr.

Lynch, Michael, and John Law. 1988. Lists, field guides, and the descriptive organization of seeing: Birdwatching as an exemplary observational activity. *Human Studies* 11: 271–303.

———. 1999. Pictures, texts, and objects: The literary language game of bird-watching. In *The Science Studies Reader*, ed. M. Biagioli, 317–41. New York: Routledge.

MacKay, Barry Kent, Paul Riss, and Matt Antonello. 2009. *A Field Guide to the Birds of Toronto.* Flap.org. www.flap.org/birdBookLongCopy.pdf.

Maguire, Joseph A. 2002. *Sport Worlds: A Sociological Perspective.* Champaign, Ill.: Human Kinetics.

Mandelbaum, Michael. 2004. *The Meaning of Sports: Why Americans Watch Baseball, Football, and Basketball and What They See When They Do.* New York: Perseus Books.

Markarian, Michael. 2005. Mute swans could find the world an alien place after protections are stripped. Humane Society of the United States. www.hsus.org/wildlife/wildlife_news/mute_swans_alien_place.html.

Markley, Robert. 2007. Climate change, techno-fixes, and systems theory. Paper presented at the annual meeting of the Society for the Study of Literature, Science, and the Arts.

Maryland Department of Natural Resources. 2003. Mute Swans in Maryland: A Statewide Management Plan, Maryland Department of Natural Resources Wildlife and Heritage Service. www.dnr.state.md.us/wildlife/msfinalstrat.html.

Mathews, F. Schuyler. 1904. *Field Book of Wild Birds and Their Music; a Description of the Character and Music of Birds, Intended to Assist in the Identification of Species Common in the United States East of the Rocky Mountains.* New York: G. P. Putnam's Sons.

Maynard, Bill. 2009. Citizen science. *Winging It, the Official Newsletter of the American Birding Association* 21 (6): 2.

McGowan, Kevin J. 2001. Frequently asked questions about crows. Cornell Lab of Ornithology. www.birds.cornell.edu/crows/crowfaq.htm.

Mech, L. David, and Luigi Boitani. 2003. *Wolves.* Chicago: University of Chicago Press.

Melley, Brian. 1997. Seagulls get reprieve; poisoning effort halted on Monomoy. www.s-t.com/daily/02-97/02-02-97.

Merriam, Florence A. 1889. *Birds through an Opera Glass.* New York: Houghton Mifflin, Chautauqua Press.

———. (published as Bailey, Florence A. M.). 1896. *A-Birding on a Bronco.* Boston: Houghton Mifflin.

———. (published as Bailey, Florence A. M.). 1898. *Birds of Village and Field: A Bird Book for Beginners.* Boston: Houghton Mifflin.

———. (published as Bailey, Florence A. M.). 1902. *Handbook of Birds of the Western United States, including the Great Plains, Great Basin, Pacific Slope, Lower Rio Grande Valley.* Boston: Houghton Mifflin.

———. (published as Bailey, Florence A. M.). 1928. *Birds of New Mexico.* Santa Fe: New Mexico Department of Game and Fish.

Miller, Olive Thorne. 1885. *Bird-ways.* Boston: Houghton Mifflin.

———. 1899. *The First Book of Birds.* Boston: Houghton Mifflin.

Miller, Thomas P. 1997. The expansion of the reading public, the standardization of educated taste and usage, and the essay as a blurred genre. In *The Formation of College English,* 30–61. Pittsburgh: University of Pittsburgh Press.

Millinery Trade Review. 1897. New York: Gallison and Hobron. www.loc.gov/exhibits/treasures/tri065.html.

Minnesota Public Radio. 2007. Bird surveys show West Nile virus devastated crows, robins, other birds in suburbia. minnesota.publicradio.org/display/web/2007/05/16/westnile.

Montlake Landfill Oversight Committee. 2002. Operational guidance for maintenance and development practices over the Montlake Landfill. www.ehs.washington.edu/epositeremed/index.shtm.

Montlake Landfill Work Group. 1999. Montlake Landfill information summary. depts.washington.edu/ubna/history.htm.

Moore, Mark P. 1993. Constructing irreconcilable conflict: The function of synecdoche in the Spotted Owl controversy. *Communication Monographs* 60 (3): 258–74.

Morris, Charles E., III, and Stephen Howard Browne, eds. 2001. *Readings on the Rhetoric of Social Protest,* 2nd ed. State College, Pa.: Strata.

Moss, Stephen. 2004. *A Bird in the Bush: A Social History of Birdwatching.* London: Aurum Press.

Multinational Monitor. 2001. The case against GE: Decades of misdeeds and wrongdoing. *Multinational Monitor* 22: 7–8. www.multinationalmonitor.org/mm2001/01july-august/julyaug01corp4.html.

Murphy, Patrick. 2000. *Farther Afield in the Study of Nature-Oriented Literature.* Charlottesville: University Press of Virginia.

Murphy, Priscilla C. 2007. *What a Book Can Do: The Publication and Reception of "Silent Spring."* Amherst: University of Massachusetts Press.

Nature Conservancy, The. No date. Interpretive trails now open at Moody Forest Natural Area. Press release. www.nature.org/wherewework/northamerica/states/georgia/press/press2183.html.

New Jersey Audubon Society 2007. Play-pledge or sponsor. worldseriesofbirding.org/wsob_pps.shtml.

New York State Wildlife Control. No date. Best practices for nuisance wildlife control operators in New York state. nwco.net/09-AppendixCBeyondTheRoutine/9-01-Crows.asp.

Nikon Corporation. 2007. Environmentally sound optical glass ("Eco-glass"). www.nikon.co.jp/main/eng/portfolio/csr/environment/products/products02/index.htm.

Norwood, Vera. 1999. Constructing gender in nature: Bird society through the eyes of John Burroughs and Florence Merriam Bailey. In *Human/Nature: Biology, Culture, and Environmental History,* ed. John P. Herron and Andrew G. Kirk, 49–62. Albuquerque: University of New Mexico Press.

Obmascik, Mark. 2004. *The Big Year: A Tale of Man, Nature, and a Fowl Obsession.* New York: Free Press.

O'Donnell, Frank. 2005. GE's greenwashing. TomPain.commonsense, May 13. www.tompaine.com/articles/2005/05/13/ges_greenwashing.php.

Orr, Oliver H. 1992. *Saving American Birds: T. Gilbert Pearson and the Founding of the Audubon Movement.* Gainesville: University Press of Florida.

Parnell, K. Harris. 2001. Toxic sludge in our communities threatening public health and our farmlands. www.toxicsaction.org/sludgereport.pdf.

Parris, Brandy. 2003. Difficult sympathy in reconstruction-era animal stories of our young folks. *Children's Literature* 31: 25–49.

Patterson, Steve. 1999. Lake Apopka: An environmental tragedy. *Florida Times Union,* www.jacksonville.com/tu-online/stories/022799/met_2a1apopk.html.

Perciasepe, Robert. 1996. U.S. Environmental Protection Agency use of the term biosolids. Memorandum. December 9. www.epa.gov/npdes/pubs/owm0132.pdf.

Perry, Matthew C., Peter C. Osenton, and Edward J. R. Lohnes. 2001. The exotic Mute Swan (*Cygnus olor*) in Chesapeake Bay, USA. USGS Patuxent Wildlife Research Center. www.pwrc.usgs.gov/resshow/perry/muteswan.htm.

Peterson, Roger Tory. 1934. *A Field Guide to the Birds,* 1st ed. Boston: Houghton Mifflin.

———. 1961. *A Field Guide to Western Birds: Field Marks of All Species Found in North America West of the 100th Meridian, with a Section on the Birds of the Hawaiian Islands,* 2nd ed. Boston: Houghton Mifflin.

———. 1980. *A Field Guide to the Birds East of the Rockies: A Completely New Guide to All the Birds of Eastern and Central North America*, 4th ed. Boston: Houghton Mifflin.

Peterson, Roger Tory, and James Fisher. 1955. *Wild America: The Record of a 30,000-Mile Journey around the Continent by a Distinguished Naturalist and His British Colleague.* New York: Mariner Books.

Peterson, Roger Tory, and Virginia Peterson. 1990. *A Field Guide to Western Birds: A Completely New Guide to Field Marks of All Species Found in North America West of the 100th Meridian and North of Mexico.* Boston: Houghton Mifflin.

Pezzullo, Phaedra. 2003. Resisting "National Breast Cancer Awareness Month": The rhetoric of counterpublics and their cultural performances. *Quarterly Journal of Speech* 89 (4): 345–65.

———. 2007. *Toxic Tourism: Rhetorics of Pollution, Travel, and Environmental Justice.* Tuscaloosa: University of Alabama Press.

Philippon, Daniel J. 1999. Introduction. In Mabel Osgood Wright's *The Friendship of Nature: A New England Chronicle of Birds and Flowers*, ed. Daniel Philippon, 1–27. Baltimore: John Hopkins University Press.

———. 2004. *Conserving Words: How American Nature Writers Shaped the Environmental Movement.* Athens: University of Georgia Press.

Popken, Randall. 1999. The pedagogical dissemination of a genre: The résumé in American business discourse textbooks, 1914–1939. *JAC: A Journal of Composition Theory* 19 (1): 91–116.

Pratt, Mary Louise. 1992. *Imperial Eyes: Travel Writing and Transculturation.* London: Routledge.

Price, Jennifer. 1999. *Flight Maps: Adventures with Nature in Modern America.* New York: Basic Books.

———. 2004. Hats off to Audubon. *Audubon Magazine*, November. magazine.audubon.org/features0412/hats.html.

Rimer, Sara. 1996. Gulls are cast as threat to avian neighbors; agency is cast in a bad light. *New York Times*, May 27.

Robbins, Chandler S., Bertel Bruun, and Herbert S. Zim. 1966. *Birds of North America: A Guide to Field Identification.* New York: Golden Press.

Roberts, Carol A. 1997. Contaminants in pelicans collected during the avian botulism event at the Salton Sea in 1996. U.S. Fish and Wildlife Service. www.fws.gov/pacific/ecoservices/envicon/pim/reports/Carlsbad/Pelican.pdf.

Rockman, Alexis. 2003. *Alexis Rockman, with Essays by Stephen Jay Gould, Jonathan Cary, David Quammen.* New York: Monacelli Press.

Rockström, Johan, Will Steffen, Kevin Noone, Åsa Persson, F. Stuart Chapin III, Eric F. Lambin, Timothy M. Lenton, Marten Scheffer, Carl Folke, Hans Joachim Schellnhuber et al. 2009. A safe operating space for humanity. *Nature* 461: 472–75.

Rosen, Jonathan. 2000. Birding at the end of nature. *New York Times Magazine*, May 21, 64–67.

———. 2008. *The Life of the Skies: Birding at the End of Nature*. New York: Picador.

Rosenbaum, Sarah. 2001. Jungle fever: Interview with Alexis Rockman. *Museo Magazine* 4. www.columbia.edu/cu/museo/Rockman.

Rothenberg, David. 2006. *Why Birds Sing: A Journey into the Mystery of Bird Song*. New York: Basic Books.

Russell, David R. 1997. Rethinking genre in school and society: An activity theory analysis. *Written Communication* 14: 504–54.

———. 2002. The kind-ness of genre: An activity theory analysis of high school teachers' perception of genre in portfolio assessment across the curriculum. In *The Rhetoric and Ideology of Genre: Strategies for Stability and Change*, ed. Richard Coe, Lorelei Lingard, and Tatiana Teslenko, 225–42. Cresskill, N.J.: Hampton.

Savage, Candace. 2007. *Crows: Encounters with the Wise Guys of the Avian World*. Vancouver: Greystone Books.

Schoch, Robert, and Michael McKinney. 1996. *Case Studies in Environmental Science*, 69–70. New York: Jones and Bartlett.

Scott, Blake. 2009. Civic engagement as risk management and public relations: What the pharmaceutical industry can teach us about service-learning. *College Composition and Communication* 61 (2): 343–66.

Scott, Mary. 1999–2002. Birding the Brownsville Dump. www.birdingamerica.com/Texas/brownsvilledump.htm.

Sheard, Kenneth. 1999. A twitch in time saves nine: Birdwatching, sport, and civilizing processes. *Sociology of Sport Journal* 16 (3): 181–205.

Sibley, David Allen. 2000. *The Sibley Guide to Birds*. New York: Knopf / Chanticleer Press.

Skitka, Linda. 2005. Patriotism or nationalism? Understanding post-September 11, 2001, flag-display behavior. *Journal of Applied Psychology* 35 (10): 1995–2011.

Slinger, Joey. 1996. *Down and Dirty Birding: From the Sublime to the Ridiculous—Here's All the Outrageous but True Stuff You've Ever Wanted to Know about North American Birds*. New York: Simon and Schuster.

Smith, G. C., and Carlisle, N. 1993. Methods for population control within a silver gull colony. *Wildlife Research* 20 (2): 219–25.

Snetsinger, Phoebe. 2003. *Birding on Borrowed Time*. Colorado Springs, Colo.: American Birding Association.

Snyder, Caroline. 2005. The dirty work of promoting "recycling" of America's sewage sludge. *International Journal of Occupational and Environmental Health* 11 (4): 415–27.

Solomon, Nancy. 2003. Birders World Series. *All Things Considered*, National Public Radio, May 11. www.npr.org/templates/story/story.php?storyId=1260267 .

Stewart, Charles J. 1997. The evolution of a revolution: Stokely Carmichael and the rhetoric of Black Power. In *Readings on the Rhetoric of Social Protest*, 2nd ed., ed. Charles E. Morris III and Stephen Howard Browne, 508–24. State College, Pa.: Strata.

Stokes, Donald, and Lillian Stokes. 1996. *Stokes Field Guide to Birds: Western Region*. Boston: Little, Brown.

Stringfellow, Kim. 2001–2009. Greetings from the Salton Sea. www.kimstringfellow.com.

Strom, Deborah, ed. 1986. *Birdwatching with American Women: A Selection of Nature Writings*, 1st ed. New York: W. W. Norton.

Swales, John. 1995. Field-guides in strange tongues: A workshop for Henry Widowson. In *Principles and Practice in Applied Linguistics: Studies in Honour of H. G. Widdowson*, ed. Guy Cook and Barbara Seidlhofer, 215–28. Oxford: Oxford University Press.

Taylor, Emily. 1850. *Conversations with the Birds*. Salem, Mass.: W and S. B. Ives.

Tice, William. 1999. *A Birder's Guide to the Sewage Ponds of Oregon: Creatures from the Brown Lagoons*. Falls City, Ore.: William Tice.

Tollefson, Jeff. 2008. Raking through sludge exposes a sting: Environmental Protection Agency scientists accused of fabricating data about health effects of fertilizer. *Nature* 453: 262–63.

Tompkins, Jane. 1985. *Sensational Designs: The Cultural Work of American Fiction, 1790–1860*. New York: Oxford University Press.

Trumbull, Gurdon. 1888. *Names and Portraits of Birds Which Interest Gunners, with Descriptions in Language Understanded by the People*. New York: Harper and Brothers.

Udvarty, Miklos D. F., and John Farrand. 1994. *National Audubon Society Field Guide to North American Birds*. New York: Knopf.

U.S. Department of Transportation. 2001. Advisory circular: Bird ingestion certification standards. www.faa.gov/regulations_policies/advisory_circu lars/index.cfm/go/document.information/documentID/22931.

U.S. Fish and Wildlife Service. 1997. Piping plovers. Informational brochure.

———. 2000. The bottom line: How healthy bird populations contribute to a healthy economy. Informational brochure. library.fws.gov/Pubs/mbd_bottom_line2.pdf.

———. 2001. Monomoy National Wildlife Refuge. Informational brochure. training.fws.gov/library/Refuges/monomoy01.pdf.

———. 2006. 50 CFR Part 10: General provisions; revised list of migratory birds; proposed rule. *Federal Register* 71 (164): 50194–50221.

———. 2007. Fact Sheet: Natural history, ecology, and history of recovery. www.fws.gov/midwest/eagle/recovery/biologue.html.

———. 2008. Bald Eagle management guidelines and conservation measures. www.fws.gov/midwest/Eagle/guidelines/bgepa.html.

U.S. Fish and Wildlife Service, compiled by John L. Trapp. 2005. What effect will the Migratory Bird Treaty Reform Act of 2004 have on bird species of your state with populations established outside their native range through human introduction? www.fws.gov/migratorybirds.

U.S. Geological Survey. 2008. North American phenology program. www.pwrc.usgs.gov/bpp.

Vermont Public Interest Research Group. 1999. On the ground: The spreading of toxic sludge in Vermont. www.vpirg.org/downloads/sludge.pdf.

Virginia Department of Game and Inland Fisheries. 2008. Small game hunting seasons. www.dgif.virginia.gov/hunting/regulations/smallgame.asp.

Washington, Sylvia Hood, Heather Goodall, and Paul Rosier, eds. 2006. *Echoes from the Poisoned Well: Global Memories of Environmental Injustice.* Lanham, Md.: Lexington Books.

Washington Department of Fish and Wildlife. 2008. Summary of general hunting season dates. wdfw.wa.gov/wlm/game/seasons.htm.

Webber, Charles Wilkins. 1855. *Wild Scenes and Song-Birds.* New York: Riker, Thorne.

Weidensaul, Scott. 2007. *Of a Feather: A Brief History of American Birding.* Orlando, Fla.: Harcourt.

Welch, Margaret. 1998. *The Book of Nature: Natural History in the United States, 1825–1875.* Boston: Northeastern University Press.

Welker, Robert H. 1955. *Birds and Men: American Birds in Science, Art, Literature, and Conservation, 1800–1900.* Cambridge: Harvard University Press.

Weller, Milton Webster. 1988. *Waterfowl in Winter: Selected Papers from Symposium and Workshop Held in Galveston.* Minneapolis: University of Minnesota Press.

Wells, Jim, Gene Aloise, Philip Olson, and Mario Zavala. 1994. Nuclear regulation: Action needed to control radioactive contamination at sewage treatment plants. GAO Report #B-255099: June 23. archive.gao.gov/t2pbat3/151920.pdf.

White, Sean, Steven Feiner, and Jason Kopylec. 2006. Virtual vouchers: Prototyping a mobile augmented reality user interface for botanical species identification. *3D User Interfaces* (3DUI'06): 119–26. www.cs.columbia.edu/~swhite/pubs/white-2006-3dui.pdf.

———. 2008. Augmented reality electronic field guide. Video. www.youtube.com/watch?v=5V9E5HlWN_k.

Wilson, Alexander, and Thomas Mayo Brewer. 1808–1814/1840. *Wilson's American Ornithology: With Notes by Jardine; to Which Is Added a Synopsis of American Birds, including Those Described by Bonaparte, Audubon, Nuttall, and Richardson.* Boston: Otis Broaders.

Witmer, Gary W., Patrick W. Burke, Will C. Pitt, and Michael L. Avery. 2007. Management of invasive vertebrates in the United States: An overview. USDA National Wildlife Research Center Symposia: Managing vertebrate invasive species. Lincoln: University of Nebraska, Lincoln.

Worster, Donald. 1994. *Nature's Economy: A History of Ecological Ideas*, 2nd ed. Cambridge: Cambridge University Press.

Wright, Mabel Osgood. 1894. *The Friendship of Nature: A New England Chronicle of Birds and Flowers*. New York: Macmillan.

———. 1895. *Birdcraft; A Field Book of Two Hundred Song, Game, and Water Birds, with Full-Page Plates Containing 128 Birds in the Natural Colours, and Other Illustrations*. New York: Macmillan.

———. 1897. *Tommy-Anne and the Three Hearts*. New York: Macmillan.

———. 1901a. *Flowers and Ferns in Their Haunts*. New York: Macmillan.

———. 1901b. *The Garden of a Commuter's Wife*. New York: Macmillan.

———. 1903. *People of the Whirlpool; from the Experience Book of a Commuter's Wife*. New York: Macmillan.

———. 1904. *The Woman Errant; Being Some Chapters from the Wonder Book of Barbara, the Commuter's Wife*. New York: Macmillan.

———. 1910. *Princess Flower Hat: A Comedy from the Perplexity Book of Barbara the Commuter's Wife*. New York: Macmillan.

———. 1922. *The Making of Birdcraft Sanctuary*. South Norwalk, Conn.: Gorham Press.

———. 1926. *My New York*. New York: Macmillan.

Wright, Mabel Osgood, Frank M. Chapman, and Ernest Thompson Seton. 1898. *Four-footed Americans and Their Kin*. New York: Macmillan.

Wright, Mabel Osgood, Elliot Coues, and Louis Agassiz Fuertes. 1897. *Citizen Bird: Scenes from Bird-Life in Plain English for Beginners*. New York: Macmillan.

Yaeger, Patricia. 2008. The death of nature and the apotheosis of trash; or, rubbish ecology. *PMLA* 123 (2): 321–39.

Zimmer, Kevin. 2000. *Birding in the American West: A Handbook*. Ithaca: Cornell University Press.

Zimmerling, J. Ryan. 2006. Why birds and birders flock to sewage lagoons. *BirdWatch Canada* 36: 4–7. www.jacquesbouvier.ca/downloads/BWCsu06.pdf.

INDEX